ORGANISMS AND ENVIRONMENTS

Harry W. Greene, Consulting Editor

1. *The View from Bald Hill: Thirty Years in an Arizona Grassland,* by Carl E. Bock and Jane H. Bock
2. *Tupai: A Field Study of Bornean Treeshrews,* by Louise H. Emmons
3. *Singing the Turtles Ashore: The Seri Art and Science of Reptiles,* by Gary Paul Nabhan

The View from Bald Hill

The publisher gratefully acknowledges
the generous contribution to this book provided by
the following organizations and individuals:

National Audubon Society Research Ranch Sanctuary
Center of the American West, University of Colorado, Boulder
Elizabeth Durein
David B. Gold Foundation
Moore Family Foundation

CARL E. BOCK
and JANE H. BOCK

WITH A FOREWORD BY
HARRY W. GREENE

The View from Bald Hill

Thirty Years in an Arizona Grassland

UNIVERSITY OF CALIFORNIA PRESS
BERKELEY LOS ANGELES LONDON

University of California Press
Berkeley and Los Angeles, California

University of California Press, Ltd.
London, England

Library of Congress Cataloging-in-Publication Data

Bock, Carl E.
The view from Bald Hill : thirty years in an Arizona grassland / Carl E. Bock and Jane H. Bock ; with a foreword by Harry W. Greene.
p. cm. — (Organisms and environments ; 1)
Includes bibliographical references.
ISBN 0-520-22183-4 (cloth : alk. paper) —
ISBN 0-520-22184-2 (paper : alk. paper)
1. Grassland ecology—Arizona—Appleton-Whittell Research Ranch Sanctuary. 2. Appleton-Whittell Research Ranch Sanctuary (Ariz.). I. Bock, Jane H. II. Title. III. Series.
QH105.A65 B63 2000
577.4′09791′79—dc21 99-047507

Manufactured in the United States of America

09 08 07 06 05 04 03 02 01 00
10 9 8 7 6 5 4 3 2 1

The paper used in this publication meets the minimum requirements of ANSI/NISO Z39.48-1992 (R 1997) (*Permanence of Paper*).

To Ariel B. Appleton
and the National Audubon Society

CONTENTS

ILLUSTRATIONS

MAPS

FIGURES

TABLES

FOREWORD

by Harry W. Greene

WITH THE PUBLICATION of Carl and Jane Bock's *The View from Bald Hill*, the University of California Press inaugurates a new series on organisms and environments. Our central themes are the distribution and abundance of organisms, the ways in which plants and animals interact with their surroundings, and the broader implications of those relationships for science and society. The series goals are to promote unusual, even unexpected connections among seemingly disparate topics, as well as to encourage books that are special by virtue of the unique perspectives and talents of their authors. Other volumes in the series include an ethnoherpetology of the Seri by Gary Paul Nabhan and a natural history of treeshrews by Louise H. Emmons.

The View from Bald Hill is about grassland ecology, written by a couple who have devoted their professional lives to untangling that mystery at a particular place in Arizona. Historically one of the largest biomes in North America, native grasslands are now among the world's most endangered ecosystems, and therein lie the most urgent implications of this fine book. Despite widespread concern for global loss of biodiversity, relatively little public attention has been focused on two of the most fundamental questions in conservation biology: How have we already affected our surroundings over the long-term past?

And just what shall we preserve for the future? If the clock could be pushed back, would we settle for North America as it existed in 1492? Mightn't we choose instead to have things as they were a few thousand years earlier, before the advent of widespread aboriginal agriculture and hunting? Or perhaps even twenty thousand years ago, before people invaded the New World? How can science help inform environmental policy, and why should we even care about the fate of prairie ecosystems?

Tucson is famous for rocky bajadas and saguaro cacti, but only a few dozen miles to the southeast is rolling grassland, punctuated by forested, outlier ranges of the Sierra Madre Occidental. Thus situated, the Bocks' study site is on the very southwestern edge of the vast plains that stretch from northern Mexico to Canada. Until the mid-nineteenth century these immense steppes sustained countless millions of insects, birds, and herbivorous mammals; those animals in turn supported numerous lizards, snakes, raptors, and other predators. Today, although cattle are widespread and large carnivores long gone, even an hour's hike or a leisurely drive in the Arizona grasslands will dispel any notions of flatland monotony. Here the incandescence of sidelit grassy hills during a thunderstorm and the orange-lavender stripes of sunset on a craggy horizon easily rival natural beauty elsewhere. In this wide-open habitat it's not unusual to have a stroll enlivened by the buzz of an agitated rattlesnake, or to spot distant pronghorn through binoculars, then realize they've long since checked you out. And it's not hard to squint at those lovely animals and drop back ten thousand years, to imagine a Pleistocene bestiary of giant ground sloths and saber-toothed cats. Later, though, as you pass through the nearby town of Sonoita, the future looms in the form of ranchettes, boutiques, and barking dogs.

The Bocks bring an unusual mix of talents to their book on this special and troubled landscape. Jane is a plant ecologist and forensic botanist, while Carl is an ornithologist; both of them are award-winning teachers at the University of Colorado. Together the Bocks have spent more than three decades studying grassland ecosystems, from South Dakota and Montana to Arizona, and they have published more than one hundred and thirty scholarly papers on the results of that work. *The View from Bald Hill* synthesizes their efforts to understand the complex effects of fire, rain, climate change, and invading species on a native biota. At the core of the story are careful observations followed by painstakingly controlled experiments, the very essence of good science, and in particular the Bocks are concerned with the effects of cattle grazing. They perceive a pattern or a puzzle in nature, then set out to learn what happens when one or more factors are manipulated; doing that can be surprisingly tedious, and often the results lead to new questions and more studies. Along the way the authors appreciate everything around them, from contrary weather and neighborly ranchers to snakes and coyotes.

Admonished to maintain objectivity and thus their credibility, biologists are also increasingly pressured to cross that line into social action, to get more involved. *The View from Bald Hill* goes well beyond a lucid account of grassland ecology and a primer on experimental field research. With graceful, unpretentious prose, the Bocks offer a firsthand view that is at once scholarly and affectionate, the chronicle of science with a heart. Above all, this book gives us a sense of the grasslands, a feel for the texture of their complexity and for their problems, and it closes with some poignant truths about our future.

PREFACE

After August with its infrared rocks,
subzero nights of large stars
have left these spikelets to flourish
as mere grace notes of straw
Decembered. Yet over our winter-killed mesas
they remain my favorite admiration.

FROM REG SANER,
"*Winter Grasses*"

THIS BOOK IS ABOUT THE Sonoita Plain, sometimes also called the Sonoita Valley, a grassland and oak savanna in the high plains of southeastern Arizona. For the better part of twenty-five summers we have worked and lived there at a place called the Appleton-Whittell Research Ranch. The purpose of this 7,800-acre property, a sanctuary of the National Audubon Society, is to serve as a biodiversity preserve and as a natural laboratory for ecological study. Our goal in this book is to describe how the biota and landscape of this extraordinary place have been changing with time.

As all naturalists have discovered, living in a wild place and trying to understand it can be a profoundly humbling experience. No doubt our research raises more questions than may ever be answered. Ecological patterns and processes reveal themselves slowly and fitfully in arid places, and then only to those with the patience to wait and to watch. After nearly three decades of study, some patterns have emerged and some changes have become evident. The time seems right to summarize our observations and speculations for the benefit of those charged with stewardship not only of this particular sanctuary, but of other places like it in the borderlands of the American Southwest.

Some hard data are included here, in the form of tables and graphs, because we believe any person interested in environmental matters deserves to see the numbers themselves and not just somebody's opinions about what they might mean. But we also have attempted to personalize our account, to make it not just informative but also interesting to anyone with a passion for the natural bounty and a concern for the future of our nation's grasslands.

This is not strictly a book about the aesthetics of nature, and we do not consider ourselves to be nature writers. Neither is it confined to the traditional boundaries of environmental reporting. There are serious conservation issues at stake in the Sonoita Valley—livestock grazing, suburban sprawl, and the spread of alien vegetation, to name the most prominent. These deserve and receive serious attention in our book. However, we find many current works on the environment to be so ponderous, heavy-handed, and gloomy that it is difficult to imagine why anybody other than a masochist or a zealot would want to bother reading them. Even so, a rosy interpretation of things is not an honest alternative, because the problems are real, and even if the paths to their resolution can be marked, it is not clear that we possess the will to follow them.

Perhaps the best way to look at the environmental mess we have gotten into—on the Sonoita Plain in particular and in the world in general—is to consider it a manifestation of the human paradox. And so it seems appropriate and comfortable to write about environmental matters with a mixture of humor and pathos, of humility and hubris, that attends to the issues themselves. This is not the usual way that scientists or environmentalists go about the business of communicating with the outside world. We are not sure if it works, but the reader deserves to know in advance that this is our goal.

The Research Ranch has had many friends—stewards, benefactors, students, colleagues, neighbors—all of whom have contributed to the security and integrity of the sanctuary and to the story we have to tell. First and foremost, Ariel Appleton and her family had the vision to conceive the idea and the generosity to give the land in the beginning. They have remained passionately involved ever since. The Whittell Foundation and the National Audubon Society provided the long-term security and support the place has needed and deserved. The United States Forest Service and Bureau of Land Management, and the Arizona State Land Department, cooperated in making public lands a part of the sanctuary and its purposes.

Neighbors of the Research Ranch, especially the Brophy and Jelks families, graciously permitted access to their lands for our cross-fence studies and shared with us their understanding and love of the land. A series of dedicated resident managers kept the fences up, the buildings and vehicles and windmills working, the budgets mostly in the black, and usually made our lives more comfortable than we had a right to expect: Larry Mitchell, Vern and Nancy Hawthorne, Mark and Barbara Stromberg, Gene and Grace Knoder, and Bill Branan.

Other friends in the Sonoita Valley have been helpful in more ways than they know: Mary Bartol, Sis and George Bradt, Paul Martin and Mary Kay O'Rourke, Joe and Helen Taylor, Mary Peace Douglas, Dorothy Sturges, and Sarah Dinham. Our daughter, Laura Bock Hernandez, accompanied us many summers in the early years. She was a loyal companion and an able, all-purpose field assistant. Finally, we could have accomplished little without the help of colleagues and students too numerous to name here, who shared in the lure of the place and helped to unravel some of its mysteries. We are grateful to all of these individuals, while we absolve them of responsibility for any of this book's factual errors or opinions they might not share.

Our fieldwork at the Research Ranch has been supported and hosted over the years by many individuals, organizations, and agencies, and we thank them all: the National Science Foundation, U.S. Forest Service, U.S. Bureau of Land Management, U.S. Army (Fort Huachuca), U.S. National Park Service, National Audubon Society, National Geographic Society, Charles A. Lindbergh Foundation, University of Colorado, Earthwatch and the Center for Field Research, Research Ranch Foundation, and Joseph and Helen Taylor.

We are grateful to those who helped us in the preparation of this book. Conrad Bahre, no stranger to the Sonoita Plain, read an early draft of the manuscript and provided many helpful comments and suggestions. Erika Geiger drew the maps and, along with Bill Branan and Bill Ervin, took some of the photographs. Finally, for their invaluable assistance, we thank Doris Kretschmer and her colleagues at the University of California Press, especially Alexandra Dahne, Dore Brown, Anne Canright, Hillary Hansen, Danielle Jatlow, and Nola Burger. It was a pleasure working with all of you.

A word about style. It is common in scientific writing to cite individuals and their publications in the text, where and when their data and ideas have been borrowed. This makes for accurate reporting but often for cumbersome reading. As an alternative, we have attempted to let each chapter flow on its own and then included notes and a literature section at the end of the book. These are designed not only to guide interested readers to more information about a particular topic, but also to give proper credit to the work of others. We hope our colleagues and collaborators accept this means of acknowledging their essential contributions to our understanding of life and times on the Sonoita Plain.

One

THE GRASSLANDS OF CORONADO

On the Sonoita Plain, beyond the far northern edge of the Sierra Madre Occidental, lies the small town of Elgin, Arizona. For many years there was a gas, grocery, and post office place in Elgin with a sign over the door that said simply, "Where the sun shines and the wind blows." The sign and the store are long gone, but it still is sunny and windy out there, especially in the dry months of May and June.

Elgin was scarcely alive in the summer of 1991, after they closed up the post office and moved the school. But its few remaining residents could see the increasing lights of Tucson and Sierra Vista against the night sky, and they probably knew that things were due for change. We moved away from the Sonoita Valley that year, and we thought it was for good.

About seven miles south of Elgin there is a rounded prominence called Bald Hill. Like most of the Sonoita Plain, Bald Hill is a high desert grassland, treeless except for an occasional mesquite and some live oak growing in washes around its base. Bald Hill is part of a rolling open valley that stretches out and then up into distant slopes of surrounding mountains (Map 1). It also is part of a unique property, managed not for cattle or grapes or housing like most of the region is today, but as a relatively undisturbed place where ecological patterns and

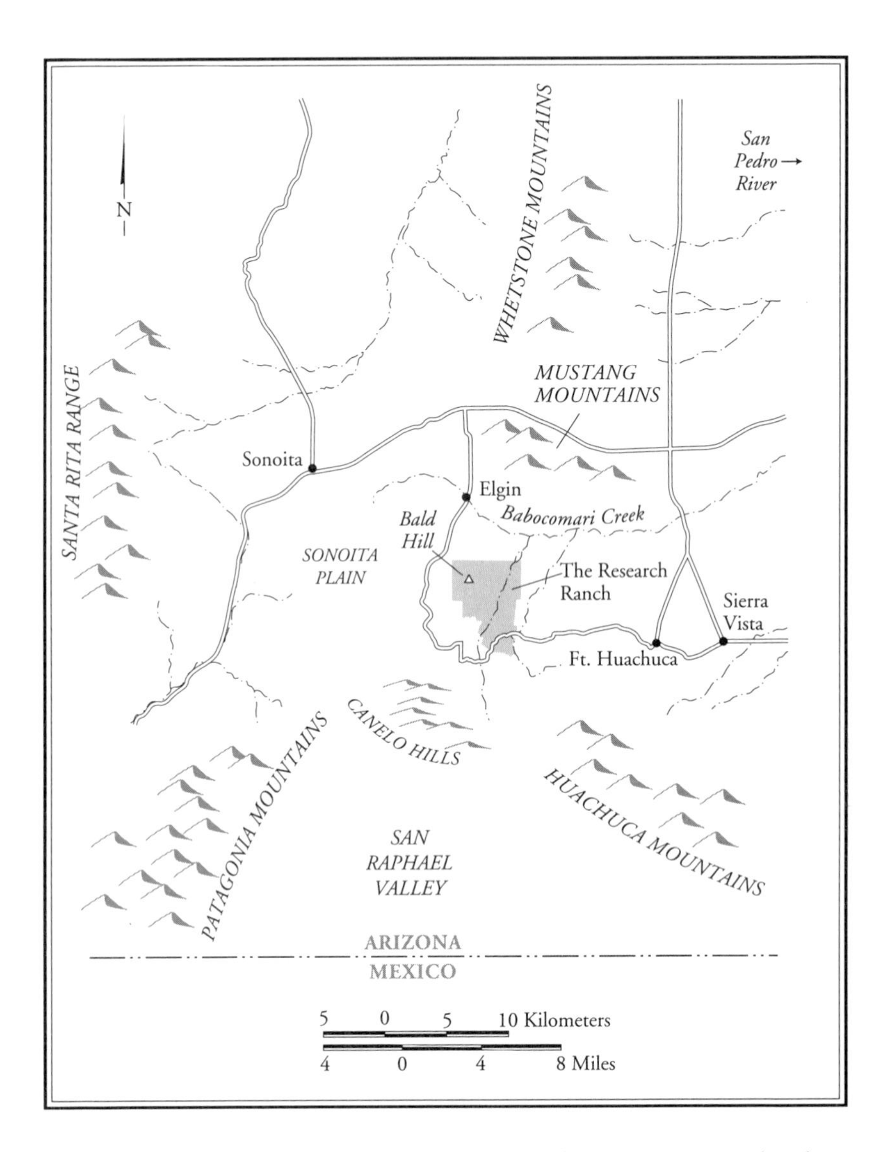

MAP 1. The Sonoita Valley and vicinity. (Redrawn from a map prepared at the Research Ranch by Erika Geiger)

FIGURE 1. The Research Ranch in 1998, looking west. Bald Hill is the low knoll to the right of the Santa Rita Range on the far horizon. (Photo by Erika Geiger)

processes in the grasslands and savannas can reveal themselves as nearly as possible without human interference. We began to study this Arizona high plain in 1974, and the present book is about some of the things we and others have learned there.

The view from Bald Hill is a classic landscape of the American West, an arid place usually more brown than green, with clear skies and far horizons (Fig. 1). It is basin and range country, like most of southeastern Arizona, except that the Sonoita Basin is higher and cooler and wetter than most, and so it is a grassland and neither Chihuahuan nor Sonoran desert.

A few miles to the north of Bald Hill are the cave-pocked limestone cliffs of the Mustang Mountains (Fig. 2). Behind the Mustangs are the Whetstone Mountains, and thirty miles east the otherworldly rocks of Texas Canyon and the Dragoons. The dark forested slopes of the Huachuca Mountains rise out of the plains to the southeast. The Santa Rita Range frames the western horizon. Even in the hottest days of

FIGURE 2. The Research Ranch in 1998, looking north to the Mustang Mountains. (Photo by Erika Geiger)

June, a patch of snow may persist atop the Santa Ritas, if it has been a wet spring. Due south, the country grades steadily and gently up into oak savannas of the Canelo Hills. Beyond these lie grasslands in the San Rafael Valley, and then Mexico—only twenty miles away as the Chihuahuan raven flies.

When the summer rains come in July or August, water runs north off the slopes of Bald Hill into Vaughn Canyon, and then east down Babocomari Creek to the San Pedro River. The San Pedro itself flows north to the Gila, which in turn feeds the Colorado. In the very old days it is possible that a drop of rain falling on Bald Hill might eventually have reached the Gulf of California by this route.

HISTORY

Prior to 1968 the history of Bald Hill was essentially the same as that of the Sonoita Plain as a whole. Humans had occupied the region for at least the preceding ten thousand years, hunting game and collecting wild plant foods and perhaps exterminating some of both. Local peoples added agriculture to their repertoire over a thousand years ago, growing crops such as maize, beans, and squash. These earliest agricultural efforts were confined largely to floodplains and low benches along watercourses. The semiarid grasslands of Bald Hill probably were not much affected.

In 1540, Francisco Vásquez de Coronado traveled north out of old Mexico, probably down the San Pedro River east of Bald Hill. His expeditionary party brought horses and cattle. Because very few if any bison had occupied southern Arizona since at least the close of the last ice age, Coronado's livestock comprised an ecological force new to the region. For the first time in thousands of years, large herds of hooved mammals grazed the grasses of the Southwest. Although other hooved mammals were common in the valley of the San Pedro and on the Sonoita Plain, including peccary, mule deer, white-tailed deer, and pronghorn, these species forage relatively infrequently on grasses. They would never have controlled the structure and function of local grasslands in the manner of bison in the Great Plains, or wildebeest in the Serengeti.

Some few livestock doubtless grazed on the Sonoita Plain and perhaps even on Bald Hill between 1540 and 1800, during which time the region was variously influenced and occupied by Spaniards, Mexicans, and Native Americans. Cattle arrived to stay in 1832, when most of the area became part of the San Ignacio del Babocomari land grant. Livestock became the dominant regional ecological force they remain to this day with the arrival of Anglo-American ranchers and the establishment of nearby Fort Huachuca, in 1877.

By the 1880s, tens of thousands of cattle were grazing the Sonoita Valley. At first the land must have seemed fertile and the range inexhaustible. Summer rains were heavy. The preceding millennia of photosynthesis without grazing had provided the perennial grass plants with vast stores of ecological capital. It proved to be a short-lived bounty, however. Cattle soon beat down the native grasses, and shrubs and mesquite trees quickly spread into formerly pure grasslands. Black-tailed prairie dogs and frequent lightning-caused wildfires had played key historic roles in keeping woody plants out of the grasslands. But the prairie dogs were extirpated, and now fires could no longer sweep across the Sonoita Valley because cattle had eaten away all the fuel.

A severe and prolonged drought finished off what little of the desert grass the cattle had spared. Most of the nineteenth century had been unusually wet in the American Southwest, and this had led to unrealistic expectations and a regional ranching mythology that nearly destroyed its grasslands. It all unraveled during the droughts of 1891 and 1892. Summer rains never came in those years, and by 1893 as many as three-quarters of the cattle had starved. Before they died, they pounded the range virtually into oblivion. This one event marked the time when the prehistoric high plains of southern Arizona were lost forever from our view.

With time the grasslands of the Sonoita Valley recovered to some degree from the drought of the 1890s, and ranching continued. While grazing intensity never again equaled that of the 1880s, it was sufficient to hold the land in a new ecological equilibrium, dominated now by certain grasses and other plant species that were more resilient to the effects of livestock grazing.

In September 1919, Juan Telles established a homestead claim around a small spring on the north slope of Bald Hill. Some time between then and now the spring pretty much stopped flowing, and Juan Telles' adobe house eroded away. Eventually, the Telles homestead became part of a much larger property, the Clark Ranch. In 1959 the Clark Ranch was purchased by Ariel and Frank Appleton (Fig. 3), who subsequently expanded the property to include the adjacent Swinging H Ranch. Together these properties made up the Elgin Hereford Ranch, which operated as such through 1967.

In 1968, a transformation occurred on the Elgin Hereford Ranch that changed the fate of Bald Hill as dramatically as had the journey of Coronado four centuries earlier. The Appletons removed all livestock and rededicated their land, now called the Research Ranch, as an environmental preserve and natural ecological laboratory. The National Audubon Society assumed management responsibilities for the property in 1980, using income from an endowment provided by the Whittell Foundation for that purpose, and the Appleton-Whittell Research Ranch Sanctuary came into being.

FIGURE 3. Ariel Appleton, with the endangered Bolson tortoise, a species she has worked to save. (Photo by the authors)

To those of us who have worked and lived there, the place has always been the Research Ranch, or TRR, or simply "the Ranch." Of course it is no longer a ranch at all, and this has been a source of understandable confusion for neighbors and colleagues. Some longtime ranching neighbors continue to use the old way of identifying a spread in the valley, by first name only. For example, the Diamond C Ranch is called simply "the Diamond C." By the same tradition they refer to the sanctuary as "the Research," and that is fine with us.

A research ranch might well be expected to be a place where ranching activities are implemented and studied. Perhaps the sanctuary, given its purposes, should be called the Not Ranch. But old and first names tend to stick, and so the Ranch probably will stay the Ranch indefinitely.

The sanctuary includes just over 7,800 acres (3,160 ha), about 40 percent of which is private land. The remaining 60 percent belongs to the people of the United States, with management responsibility being about equally divided between Coronado National Forest and the Bureau of Land Management. Through cooperative agreements between the National Audubon Society and these agencies, the former federal grazing leases that are part of the Appleton-Whittell Research Ranch have been dedicated to the overall research and conservation goals of the sanctuary.

THE LAY OF THE LAND

For more than twenty-five years we have been involved in conducting and coordinating field studies at the Research Ranch, with the goal of understanding the dynamics of its ecosystems, the natural history of its flora and fauna, and impacts of human land use in the Sonoita Plain as a whole. After all that time, the landscape of the Ranch and vicinity has become as familiar as the lines on an old friend's face. We can talk and plan our work using a kind of shorthand geography that is as useful to us as it would be meaningless to strangers. Because the stories we want to tell in this book happened in real places—and the nature of those places influenced the outcome of the stories—it is necessary at the outset for the reader to gain some appreciation for the geography of our study area.

Elevations of the sanctuary range from 4,600 to 5,100 feet (1400 to 1560 m), the high points being Bald Hill in the northwest and ridges above Lyle Canyon in the southeast (Map 2). Three major drainages transect the property, each flowing from south to north as tributaries of Babocomari Creek (see Map 1). Vaughn and Lyle Canyons cut through the northwestern and southeastern corners of the sanctuary, respectively. The major drainage is O'Donnell Canyon, which receives

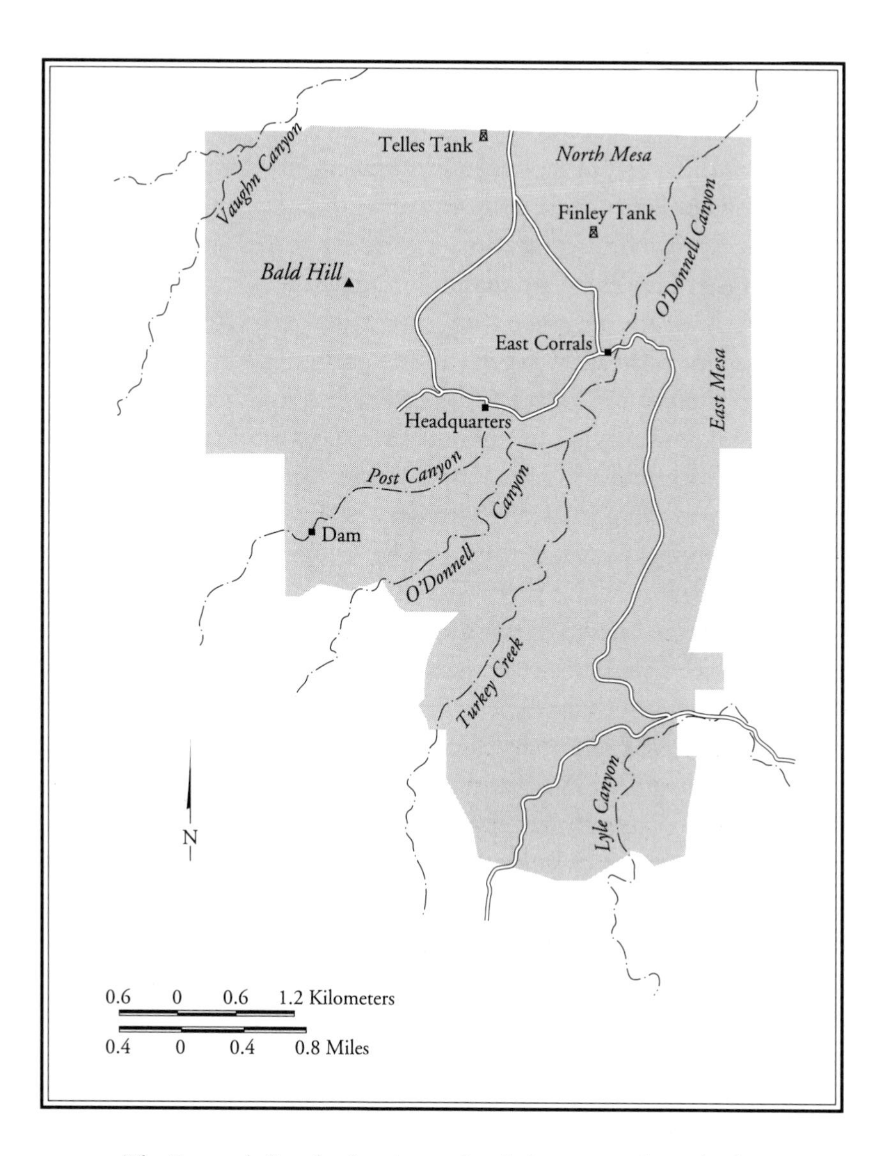

MAP 2. The Research Ranch, showing major drainages, roads, and other features. (Redrawn from a map prepared by Erika Geiger)

two tributaries centrally on the sanctuary, Turkey Creek from the south and Post Creek from the southwest, and then flows northeast off the Ranch into Babocomari Creek. All of these drainages may have had year-round flowing water historically, but for the most part today the streams run only seasonally.

Two very different sorts of vegetation follow the drainages. The first is riparian woodland, consisting of varying mixtures of sycamore, walnut, ash, cottonwood, willow, live oak, and assorted shrubs. The second is a broad floodplain grassland dominated by sacaton, a tall grass that grows in nearly monocultural stands wherever stream drainages are broad and gentle enough for accumulation of deep soil sediments.

Uplands include Bald Hill and the North Mesa, between O'Donnell and Vaughn Canyons in the northern and western portions of the sanctuary. The broad and level East Mesa lies between O'Donnell and Lyle Canyons along the eastern boundary of the property. To the south are more heavily dissected ridges between Post, upper O'Donnell, Turkey Creek, and Lyle Canyons.

Upland vegetation on the Research Ranch consists of relatively open grasslands with scattered small mesquite trees, especially in the northern half of the sanctuary. These grade southward into increasingly wooded savannas of Emory oak and Arizona white oak. Dominant upland ground cover includes short and mid-height perennial grasses, especially grama grasses of the genus *Bouteloua,* threeawns (*Aristida*), and lovegrasses (*Eragrostis*). On localized limestone outcrops there also are some plants with a strong affinity to the Chihuahuan Desert: ocotillo (a relative of the famous boojum tree of Baja California, so named after Lewis Carroll's fanciful poem "The Hunting of the Snark"), a white zinnea, some acacia shrubs, and grasses such as black grama that are uncommon elsewhere on the sanctuary.

There are two very important permanent water sources on the Ranch. One is Finley Tank, an impounded perennial spring in a ravine off the North Mesa. The second is a dam on Post Canyon that holds enough water even in the driest years and months to support a small cattail wetland.

Two facilities provide living and work space for researchers and the resident manager of the Ranch. These are Headquarters (the original

Clark Ranch buildings) and East Corrals (the original Swinging H Ranch). We lived most summers, and a large part of many winters, in a house at East Corrals with a fine view looking north down O'Donnell Canyon and on toward the Mustang Mountains. To us, there was something mysterious and alluring about those dry hills, and if the Mustangs appear too many times in the photographs we have included in this book, that is why.

Two

ON BEING THE CONTROL

In any well-designed scientific experiment, there must always be a control. For field research, the control usually is an unmanipulated site, against which a manipulated experimental site can be compared. Without the control site, we have no way of knowing for sure if the manipulated site changed as a result of the experiment or for some other reason. Many of the stories we have to tell about the Research Ranch have to do with its history and continuing function as a control. What follows is one example.

Burroweed (*Isocoma tenuisecta*) is a small native shrub that in certain places in the Sonoita Valley is very common. Woody plants sometimes become more abundant in grazed grasslands because they have easier access to nutrients and sunlight and water when competition from grasses is reduced. It would be reasonable to hypothesize that abundance of burroweed has increased as a consequence of historic livestock grazing in the region, although it probably always has been present at some level. How would one gather data to test such a hypothesis? One obvious way would be to compare abundances of burroweed between areas that have a long history of grazing, versus livestock exclosures that have never been grazed. Unfortunately there are no such livestock exclosures, either in the Sonoita Valley or scarcely anywhere else in the

American West. Like so many human activities, western ranching has been a huge ecological experiment largely without a control. As a result it is very difficult to quantify exactly the historical and current environmental impacts of livestock grazing.

In the absence of any livestock exclosures built by Coronado or his contemporaries—an implausible scenario to begin with—probably the next best approach to making grassland ecology a true science is to establish new exclosures on formerly grazed land and then to study these. They can serve as one sort of control site, against which the consequences of grazing or any other land use can be compared. There are a number of such exclosures throughout the West, but most of them are little more than tiny ungrazed islands in a vast sea of livestock. One larger area that has not been grazed significantly in many years is nearby Fort Huachuca, but past and continuing military activities limit the value of this site as a long-term control. The Research Ranch is another place that may be sufficiently large to reveal the dynamics of whole intact ecosystems when they are left ungrazed, and where other sorts of human disturbance are kept to the absolute minimum. Certainly that was our prime motive for studying the sanctuary in the first place.

We have measured the density of burroweed at a number of places and times on the Research Ranch and on adjacent grazed land (Fig. 4). In 1982 there were fewer than 5 plants per 100 square meters of ground, either on or off the sanctuary. By 1985 densities had risen to more than 40 plants per 100 m^2 on both sides of the boundary fences. The intervening years had been wet ones, especially in winter, and we suspect that germination of seeds and survival of burroweed seedlings may have depended upon this above-average rainfall. By 1990 many of the plants had died and densities were down to about 20 plants per 100 m^2, again on both sides of the fences. In 1995, densities were back up again, apparently in response to another period of relatively wet winters.

A final observation we and others have made about burroweed is that it is very sensitive to fire. Over the years wildfires on the Ranch have resulted in virtually 100 percent burroweed mortality in many places.

The picture that emerges about burroweed is one of a short-lived woody plant, negatively impacted by fire but able to recruit new seedlings at a rapid rate in years following wet winters. Our results in

FIGURE 4. Erika Geiger measuring a burroweed shrub on the North Mesa in 1999. (Photo by the authors)

this particular case illustrate both the power and the limitations of the Ranch as an ecological control site. Without data from the sanctuary we might have concluded, erroneously, that increases in burroweed populations in the Sonoita Valley between 1982 and 1985 were the result of grazing. With data from the Ranch as well as grazed areas, we are now certain that the current population dynamics of this shrub are much more related to fire and weather than to livestock activities.

Does this mean that the introduction of cattle into the Sonoita Valley had no influence on burroweed abundance? Not necessarily. We know that livestock reduce the frequency of wildfire by eliminating most burnable fuels and thereby curtail a major mortality factor for this particular shrub. It also is possible that cattle and horses facilitated the initial establishment and spread of burroweed, by reducing the abundance of competing grasses. Recall that the whole area was devastated by grazing and drought in the 1890s and that some level of grazing continued on what was to become the sanctuary until 1968. It is likely that these impacts were so prolonged and so profound that post-grazing ecological change is far from complete. In subsequent chapters, we consider what the sanctuary is telling us about the impacts of grazing on other sorts of plants and animals that live in the Sonoita Valley.

The limitation of the Research Ranch is that it cannot tell us with certainty what the Sonoita Plain was like before the time of Coronado, nor what it may be like one hundred years from now. The challenge to those of us who study the reserve is to interpret ecological patterns and processes as we see them today and to read them as possible clues to both the past and the future. One growing body of literature suggests that grassland ecosystems can exist in multiple stable states and that recovery from a major disturbance need not always result in a return to previous conditions. This may well prove to be the case in the Sonoita Valley, where losses of topsoil associated with drought and overgrazing in the late 1800s may never permit the grasses to flourish as they once did.

One thing is certain. With every passing year since the sanctuary became a control area, we have learned more about the potential of this land to support and sustain the endemic flora and fauna, about the spatial and temporal dynamics of its ecosystems, and about the ecological glue that holds these things together.

A final observation about being a control area is that it can be a decidedly nonglamorous business. Marking time by watching the grass grow requires both money and patience, and sometimes it has been hard to find enough of either.

There probably are few if any behavioral traits that are both unique to humans and ubiquitous among them, but one possible candidate is the desire to fiddle with things and change them for the better. We have been exhorted over the years to apply all sorts of more active methods of "improving" the sanctuary, and there is a large body of range management literature that describes ways to do it. Fertilizing, bulldozing, root plowing, chaining, mowing, shredding, contour furrowing, water-spreading, applying pesticides, reseeding with native or exotic grasses, grazing in short-duration rotations, and churning up the soil with the hooves of excited cattle—all of these may well cause grasslands to change, and perhaps in ways considered by some people as beneficial. But the very purpose of a place like the Research Ranch is to serve as a control, against which the consequences of all these sorts of intrusive manipulations can be compared and therefore better understood.

Certain ranchers and range managers have predicted that grasslands on the Research Ranch will degrade over time without some sort of essential stimulation provided by livestock. Perhaps they are right. Our goal is to monitor post-grazing ecological changes, whatever they are. If the sanctuary eventually becomes as barren as a strip mall parking lot, then our studies will have documented the importance of livestock in sustaining the southwestern plains.

But don't bet on it.

Three

WAITING FOR RAIN

THE RHYTHMS OF LIFE ARE driven by climate in the Sonoita Valley—mostly warm but sometimes frozen cold, mostly arid but sometimes soaking wet. Wild things have evolved to tolerate such extreme and unpredictable changes in their environment, otherwise their ancestors would long since have gone extinct. Humans, or at least those living there now, have a much more difficult time of it. We have never lived in another place where people are so obsessed by the weather. This is especially the case in the heat of June, when the parched land and everybody on it awaits arrival of the summer rains. To be sure, part of the obsession is economic, since the ranchers and their livestock depend on rain. But ranchers are an ever-decreasing proportion of human residents in the Sonoita Valley. Nonetheless, the obsession persists as strongly and universally as ever. Will the rains be early or late? Will they be strong? Will they come at all?

Waiting for rain on the Sonoita Plain is a dusty, hot, windy endurance test, and one's patience can wear thin. This is particularly true if the wet season starts feebly. There will be just a few small thunderclouds around, and it always seems as if God is watering somebody else's grass. Everyone has a rain gauge, and comparing daily amounts in the smallest fractions of inches is the predominant social interchange. A rancher once confessed to us that he liked it when flies and bees drowned in his rain gauge because it made the water line go up and so increased his boasting rights.

Conversations like the following, which must seem cryptic to an outsider, are common in local bars in July and August:

"Did you get any last night?"

"About thirty-five hundredths. Did you get any?"

"Hell no."

Two sorts of temporal climatic variation have had profound effects on life at the Research Ranch. The first is variation between months of any given year, and the second is variation among whole years. We shall consider these in turn.

THE SEASONS

Four great air masses, each distinctly different from the others, compete for space over the basin and range country of southeastern Arizona, flowing in from the equator, the Arctic, the Pacific, or the Gulfs of California and Mexico. Their interactions determine the seasons of life in grasslands and oak savannas of the Sonoita Valley (Fig. 5).

The sun's warming effect is greatest near the equator, causing tropical air to rise and then to spread toward both poles. This air descends back toward earth at about thirty degrees north and south of the equator. Because air becomes both warmer and drier as it falls, these are the latitudes of the world's hottest and most arid grasslands and deserts, including those of the American Southwest. Unless otherwise displaced, this descending, dry, once-tropical air dominates the Sonoita Valley, imparting to the region its fundamental aridity.

In winter, moist storms sweep onshore along the Pacific slope of North America, bringing their characteristic cool-season rains to Washington, Oregon, and California. In the south, the strongest of these storms may push across into Arizona, temporarily displacing the desert air and bringing winter moisture to the region. For the most part these are gentle storms that come between December and March, with flat

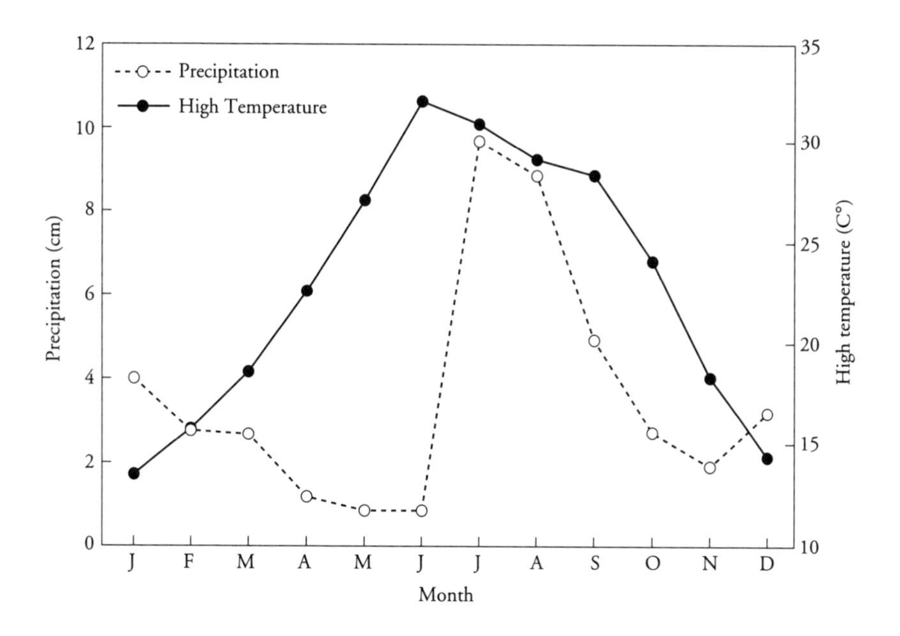

FIGURE 5. Mean monthly precipitation and mean daily high temperature at the Research Ranch. Data are from the sanctuary weather station for the years 1968–1996 (rainfall) and 1981–1992 (temperature).

gray skies and daylong drizzle. Occasionally, however, Arctic air drives south into Arizona, mixes with moist Pacific air, and brings snow to the Sonoita Valley. Snow and cold eliminate any plant or animal whose tropical origins preclude tolerance to such severe winter weather.

Pacific and Arctic air masses retreat northward beginning in April, and the Southwest is hot and dry in May and June. In the lower desert, early spring can be a period of abundant life, when plants respond to the combination of warmer air and soil moisture stored from winter rains. But except for a few kinds of European weeds that are adapted to late-winter rains and cool weather, the spring bloom is scarcely noticeable at higher elevations such as the Sonoita Valley. Here, the land remains too cool to allow active growth until well past the end of the winter rainy season.

June is the bleakest month at the Research Ranch—what Ariel Appleton once described to us as the "Dead Season." It usually is bone dry, and last year's grass just lies there, disheveled and gray, beaten by the dry spring winds and bleached by the white light of the desert sun. The land and most of its inhabitants are on hold, waiting for the arrival of the fourth great air mass.

(When the National Audubon Society first considered buying the Research Ranch, back in the mid-1970s, somebody in the New York office decided it would be a great idea if the national board of directors came for a tour—in June when, they were sure, the place would be at its most beautiful. We reacted with horror. "No," we said, "please come in August when the grass is green." They said, "What?" And they came in June anyway.)

Just when it seems as if an uninterrupted pale blue sky will forever dominate the Sonoita Plain, one afternoon in early July a puffy white cloud will form over the Huachucas or the Santa Ritas. It will not rain on that day, and perhaps not for a week or more, but the cloud has signaled the arrival of new air, coming up from the Gulf of Mexico and from the Gulf of California and circling far enough north and west to touch the desert grass. This is the summer monsoon.

Monsoon rains in the Sonoita Valley are vastly different from the dreary storms of winter. They come as great convective thunderstorms that form over the mountains and then sweep out over the grasslands. These summer storms are dramatic, drenching, and often violent, but usually localized. When a watershed upstream from the Ranch fills to overflowing, you can hear the rumble of the water five miles away as it moves toward and then on and through the sanctuary.

Lightning-caused wildfires are at their peak just before the start of the summer rainy season, when the grass is still brown. Fire plays a number of vital roles in the dynamics of the Arizona high plains, as we shall see. But if it is a good year, by late July the grasslands of the Research Ranch will be as green as a Welsh countryside, and they will remain so until the killing frosts of November.

In grasslands on the Great Plains of North America, May and June are both wet and warm, and so the wake-up call of spring is an unambiguous event. Not so in southeastern Arizona, where the hot days of spring and early summer are also the driest. Plants and animals of the Sonoita Valley therefore receive decidedly mixed signals as to when they should begin their annual business of growing and reproducing.

There are, in fact, two springs, an early one that is thermal and a later one that is hydrologic.

To which spring should an organism respond? We have found that the answer depends very much on habitat. Trees and shrubs of the riparian communities and oak savannas usually begin to grow new leaves with the arrival of thermal spring in May and June, perhaps because their roots have access to groundwater. Animals of these habitats also become active early. Characteristic woodland birds, such as the bridled titmouse, ash-throated flycatcher, Lucy's warbler, and Bewick's wren, mostly have finished breeding by mid-July.

Grasslands of the Research Ranch usually do not green up until late July, and do not reach their peak production until mid-August. Therefore, most animals of grassland habitats respond to hydrologic spring, in July. Birds such as Cassin's sparrows, grasshopper sparrows, and eastern meadowlarks nest from late June through to early September, with August being the peak breeding period.

YEARLY VARIATION

In the winter of 1978–1979, a wave of unusually cold Arctic air reached south to the Sonoita Valley. For the better part of a week, nightly temperatures fell far below freezing, on several occasions dipping below zero degrees Fahrenheit. This must have been one of the coldest episodes of the twentieth century in southeastern Arizona. Plumbing and water supply systems that had been parts of old ranch buildings for fifty years or longer froze and burst. Oak and mesquite trees that were hundreds of years old were killed to the ground.

One animal that suffered horribly from the freezes of that winter was the collared peccary, also called the javelina. For several years after the freeze we scarcely could find any javelinas on the Ranch at all. This species extends well down into South America and has strong tropical roots. It lacks the heavy insulative body fur and fat of mammals that evolved in colder climates.

We are certain that most of our javelinas perished from the cold in that unusual winter. However, by the mid-1980s javelinas were once again common on the sanctuary. We do not know if they spread back up from the tropics, from the low desert, or from local refugia, but their return illustrated a fundamental ecological principle. All species have the reproductive ability to quickly exceed the carrying capacity of prime habitat. This generates pressure on younger individuals to disperse away from their natal areas, to escape the competition from others of their own kind. In this way a species is always straining against its distributional limits, pulsing outward at times when the environment is benign, only to be repelled when the environment again becomes harsh.

Even more than winter cold, strength of the summer monsoon is a variable that shapes the dynamics of Sonoita Valley grassland ecosystems. We have already noted how the droughts of the early 1890s, coupled with livestock overgrazing, had a devastating impact on the flora and fauna of the region and on the livestock industry itself. While there have been no droughts of similar intensity since we have begun working at the Ranch, nevertheless some years were much drier than others (Fig. 6). Between 1968 and 1996, annual rainfall extremes were a high of 28.7 inches (72.9 cm) in 1983 and a low of 10.3 inches (26.2 cm) in 1996, compared to the twenty-nine-year average of 17.1 inches (43.5 cm). During the critical monsoon months of July and August, the twenty-nine-year low was 3.2 inches (8.1 cm) in 1979, while the high was 12.7 inches (32.2 cm) in 1990.

These annual differences in rainfall have had profound effects upon the flora and fauna of the Research Ranch. For example, seed production by grasses and herbs was much lower following dry summers, which in turn resulted in smaller wintering populations of migratory seed-eating sparrows. Another example is plains lovegrass, an important perennial bunchgrass that thrived and increased in wet years but suffered considerable mortality following dry years. Long-term abundance of this key grass species in the Sonoita Valley depended on powerful interactions between rainfall, fire, and livestock grazing, as we shall describe later.

A major and developing issue in the American Southwest is the relative strength of the summer versus winter rains, how this affects the abundance of grasses versus woody plants, and how it all may be influenced by possible human-induced global warming and climate

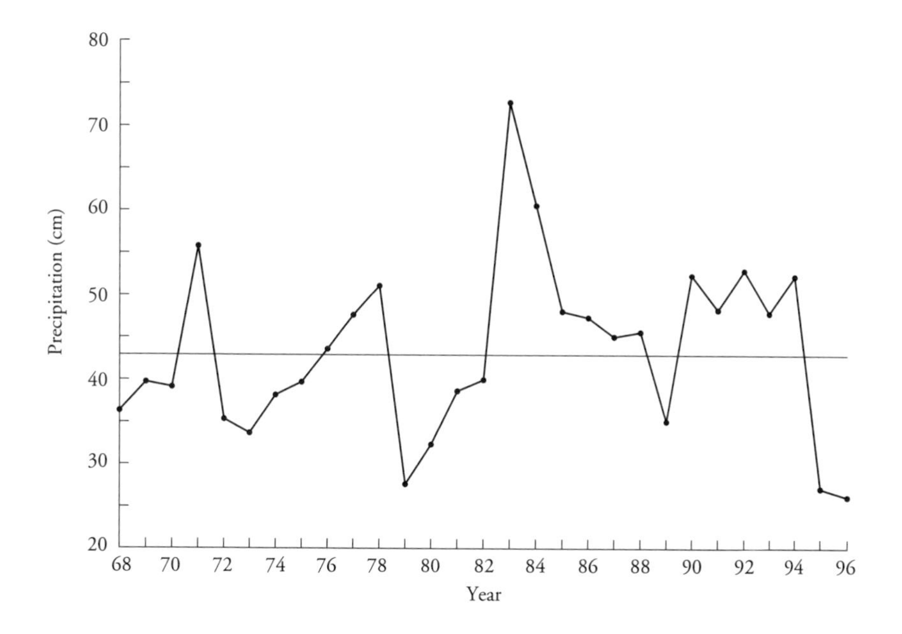

FIGURE 6. Total annual precipitation at the Research Ranch, 1968–1996. Horizontal line is the twenty-nine-year mean.

change. There is good evidence that woody plants such as burroweed and mesquite are better able to take advantage of winter rains, while southwestern perennial grasses are dependent upon the summer monsoon. Historical and paleoecological evidence makes it pretty clear that episodic shifts from desert grassland to desert shrubland, and back, have tracked changes in the relative strengths of summer versus winter rains, and that livestock grazing has greatly favored the first sort of change over the second. It remains to be seen how global climate change might influence the delicate balance between desert grass and desert shrub, though some have predicted that it, too, may favor the woody species. In any event, control areas such as the Research Ranch should prove vital as places to measure the impacts of climate change in the comparative absence of other, confounding, human influences.

For all its harshness, the Dead Season in southeastern Arizona is a ritual to be survived and appreciated. The first cooling rain and the pungent smell it makes in the dust are all the sweeter for the wait. It is not possible to understand or appreciate life in the Sonoita Valley without knowing the spring drought first-hand. The late Virginia Rich, an author and area resident, once told us she didn't think people who went away for June deserved to come back in July.

The summer of 1983 was unusually wet. One day in August we were over in Lyle Canyon collecting plant cover data on some plots that had burned the previous year. The grass was as tall and lush as we could recall. It was hot, and the air was still and very humid. Sweat poured off our faces and arms as we swatted in vain at the clouds of mosquitoes and no-see-ums around us. All the draws and washes on the Research Ranch were running full, and we hoped to avoid getting stuck in the mud for the second time that day when it came time to head home.

A ranching neighbor came by and stopped to pass the time. The conversation naturally turned to the weather and the grass, and we all agreed that the Sonoita Plain looked fine. "You know," he said, "I've lived in this country for twenty-five years, and this is about the first typical summer we've had."

In retrospect, we all tend to remember the monsoons that were above average, when the grasslands were at their most grand. We tend to forget the droughts. That is why there is a local mythology that it never rains like it used to.

Four

FENCELINES

MOST RANGELAND FENCES in the Sonoita Valley were built in the 1930s, out of barbed wire and wooden posts made from the trunks of alligator junipers, which are particularly resistant to decay. Foothills surrounding the valley probably were locally depleted of junipers, so great was their harvest by the builders of these original fences. Since those days, waxwings and solitaires and bluebirds have helped to replenish junipers all over the region, by eating their cones and dispersing their seeds. One irony is that lots of new young junipers are growing next to the steel posts that have largely replaced their ancestors, because that is one place where the birds go to perch and do their metabolic business.

Most of the original fences in the Sonoita Valley had a clear and simple purpose—to keep one rancher's cattle from eating another rancher's grass. Fences surrounding the Research Ranch play a more complicated role, with an environmental significance extending beyond that of their predecessors. We have been there when ranchers pointed across the sanctuary's fences and pronounced its grasslands "decadent." We have been there when visiting environmentalists stood at those same fences and, pointing the other direction, condemned ranching as the number one evil in the West.

If studies at the Research Ranch are to be effective and influential, then the sanctuary's role as an ecological reference point must not be weighted down by the cultural and political baggage presently attending much of the debate about livestock grazing. The central question of the present chapter is this: After thirty years without livestock, how have grasslands on the sanctuary changed compared to those of adjacent operating cattle ranches? We attempt to answer this question, tentatively and speculatively to be sure, but as objectively as we can, without using the value-laden terminology that frequently has turned the grazing debate into a shouting match.

THE ISSUE

Of all the economic uses of western public land, none is more widespread than livestock grazing, and none more controversial. Part of the controversy stems from the different agendas of the various parties involved. Ranchers, land managers, range scientists, recreationists, conservationists, and the tax-paying landowners have distinctive perspectives on the purposes of public land.

What makes the livestock grazing issue particularly vexing, however, is the ambiguity about its actual environmental consequences. Some level of grazing by mammals is a natural part of most grassland ecosystems, and it may be that cattle are simply substituting for native grazers. On the other hand, not all grasslands have had equal evolutionary associations with large grazing mammals, and in any event, the activities of fenced, predator-proofed, domestic grazers are unlikely to be the same as those of their wild relatives.

When a cow bites the top off a grass plant, the stems and leaves usually grow back. Regrowth may not occur until the next rain or the next growing season, but it almost always will happen within a year. In some cases, the grass plant may grow back more vigorously than before, because the cow has cleared away the old dead stems and converted them to manure that can fertilize the range. It can and frequently has been argued that moderate grazing has no permanent effects, or is perhaps even beneficial. Most grass plants in fact have considerable material and energy stored in their root systems, and usually they respond to clipping by rapidly moving some of both from roots to shoots. If grass plants did not do this, it would not be necessary to mow the lawn nearly as often.

Contrast this with what happens when a logger cuts down a stand of three-hundred-year-old Douglas fir. No one can dispute that the trees are now gone, or that the loss is essentially permanent on a time scale relevant to humans. The effects of livestock on grasslands are much more subtle than the effects of logging on forests, and assessing the consequences of grazing requires careful measurement over long time periods and under many circumstances. However, the effects of grazing can be no less important than those of logging when they are viewed from the perspective of particular grasses in specific sorts of ecosystems and over sufficient time.

THE POTENTIAL VALUE OF THE RESEARCH RANCH

As we saw in Chapter 2, one problem with the virtual ubiquity of livestock grazing on western American rangelands is that the herds have performed a huge ecological experiment largely without a control. Most of this continent's grasslands were colonized by Europeans and their livestock so thoroughly, and so long ago, that we have no sure idea what they looked like or how they functioned in pre-Columbian times. The challenge facing conservationists and grassland ecologists is to learn as much as possible about the historical condition of these grasslands in order to craft realistic and defensible strategies for their conservation or possibly even their restoration.

The livestock grazing issue has been central to so much of our work at the Ranch that it is difficult to write about it without sounding self-important. For this reason, it seems like a good idea to remind the reader what field research on the sanctuary can, as well as what it cannot, tell us about life on grasslands in the Sonoita Valley and elsewhere in the West.

First, there is the good news. Certainly the Ranch, at 7,800 acres, is large enough to contain viable populations of all but the most wide-ranging top predators. In addition, there has always been a resident manager present to repair leaky fences, and the neighbors have been good about retrieving their livestock when necessary. These same ranching neighbors also have been unfailingly generous in granting us permission to make comparative studies across the sanctuary boundary fences. Another value of the Ranch is that it has been relatively well studied compared to many exclosures. A large data base has been accumulated, quantifying the nature of the sanctuary, how this has changed since livestock were first removed in 1968, and how it now differs from adjacent grazed areas.

It is equally important to recognize and acknowledge the limitations of the Ranch. First, while data were gathered on the conditions of the sanctuary lands at the time cattle were first removed in 1968 and 1969, no comparable information was collected in those early years on the condition of adjacent ranches that have continued to be grazed. We can therefore attribute present-day cross-fence differences to livestock exclusion on the sanctuary only if we assume that these differences did not exist prior to 1968. This is a reasonable assumption, given that the pre-1968 land-use history of the sanctuary as a working cattle ranch was similar to that of the Sonoita Valley as a whole, but it is an assumption nevertheless.

Second, the lessons of the Research Ranch must not be applied too widely across the West. It is but one small place in a large and diverse array of western grassland ecosystems. Different sorts of grasslands respond differently to livestock, or to livestock exclusion, depending largely upon climate, soils, and their historical association with native grazers. Two examples illustrate this point.

Certain very arid grasslands of the Southwest were converted to desert scrublands by grazing, probably coupled with subtle changes in climate, at the end of the last century. Livestock exclusion today has little effect on these lands, apparently because soil changes associated with historic overgrazing have rendered them permanently incapable of supporting much in the way of grasses.

At the other extreme, many parts of the western Great Plains had such prolonged and powerful associations with bison that most of their grasses evolved high tolerances for the activities of large grazers, native or domestic. Livestock may have affected these ecosystems relatively little, or may even have helped sustain them in something like their prehistoric condition. Livestock exclosure in the twenty-first century may not result in much change, since there are few grazing-intolerant plants or animals around now. Lack of grazing may even cause the decline or disappearance of species native to these grasslands, particularly those animals such as prairie dogs and certain nesting birds that require relatively short and sparse vegetation.

We believe the lessons of the Research Ranch are applicable most specifically to surviving semiarid grasslands in the Southwest and Intermountain West, where native large grazers were scarce or absent.

Finally, we must remember that there is no historical record describing the nature of the Sonoita Valley prior to the time of Coronado. We do see changes happening on the grasslands of the Research Ranch that almost certainly are attributable to livestock exclusion. By projecting the direction and speed of those changes into the future, we may be able to predict how the land will look in fifty or one hundred years. It requires another bold step to assume these changes are telling us anything about the past. However, we cannot resist taking that step.

Is the ecological clock running backwards at the Research Ranch, marking a time before the extreme overgrazing of the late nineteenth century? We can never be sure, but one sort of circumstantial evidence seems strongly to favor such a notion: namely, the abundance and variety of native plants and animals that today are increasing on the Ranch and that otherwise are relatively uncommon in the Sonoita Valley. The very presence of these grazing-intolerant species is powerful testimony that ecological and evolutionary opportunities must have existed for them in the not-too-distant past.

THE CAST OF CHARACTERS

Nineteen different perennial grass species dominate grasslands of the sanctuary and surrounding cattle ranches. During the summer of 1983 we quantified the basal-area ground cover of each of these grasses on 132 twelve-meter-diameter plots scattered widely over the northern two-thirds of the sanctuary. Basal-area ground cover is the percentage of the ground occupied by a grass plant at the junction between its above- and belowground parts.

TABLE 1. Percent basal-area ground cover of dominant perennial grasses in various habitats on the Research Ranch

Species	*Floodplains and Washes*	*Low Benches*	*Level Uplands*	*Rolling Uplands*	*North-facing Slopes*	*South-facing Slopes*
Blue grama (*Bouteloua gracilis*)	23.9	21.6	15.4	2.6	5.9	3.6
Hairy grama (*Bouteloua hirsuta*)	0	1.1	0.5	1.8	1.7	1.9
Sideoats grama (*Bouteloua curtipendula*)	5.7	6.4	1.7	1.3	10.8	9.7
Sprucetop grama (*Bouteloua chondrosioides*)	0	1.8	2.4	6.3	1.3	4.5
Black grama (*Bouteloua eriopoda*)	1.3	0.5	0.6	1.7	1.3	2.9
Plains lovegrass (*Eragrostis intermedia*)	0.6	2.2	5.3	7.2	2.8	2.4
Arizona cottontop (*Digitaria arizonica*)	4.9	0	0.6	0	0	1.4
Curly mesquite (*Hilaria belangeri*)	0	0.7	0.8	11.7	5.2	9.1
Wolftail (*Lycurus phleoides*)	0	0.4	4.3	3.1	1.2	0.5
Threeawns (*Aristida* species)	0.2	0.4	5.3	2.0	4.6	2.0
Vine mesquite (*Panicum obstusum*)	4.5	1.3	2.4	0	0	0
Big sacaton (*Sporobolus wrightii*)	6.0	0.2	0	0	0	0
Cane beardgrass (*Bothriochloa barbinodis*)	0.3	0	1.8	0.2	0.1	0.5
Texas beardgrass (*Schizachyrium cirratus*)	0	0	0.4	0	2.1	0.1
Tanglehead (*Heteropogon contortus*)	0.5	0.2	0	0	0	1.9

Data are from 132 12m-diameter circular plots sampled in August 1983.

During the study we recognized six habitats as distinctive, marked by their dominant grasses (Table 1):

1. floodplains, dominated by blue grama, sacaton, and sometimes vine mesquite or Arizona cottontop
2. low and level benches, lying just above floodplains, usually dominated by blue grama or by a combination of blue grama and sideoats grama
3. level upland mesas, with grass cover mostly of blue grama, plains lovegrass, wolftail, and the different threeawn grasses
4. rolling uplands, usually rockier than mesa tops, and dominated by plains lovegrass, curly mesquite, and sprucetop grama
5. north-facing slopes, often with a partial canopy of oak trees, where common grasses are sideoats grama, blue grama, plains lovegrass, and occasionally Texas beardgrass
6. south-facing slopes, with highest ground cover of sideoats grama, curly mesquite, sprucetop grama, and, especially on limestone outcrops, black grama

This rather detailed grassland habitat analysis makes clear the spatial heterogeneity of the Ranch. Because of this complexity, our cross-fence comparisons of grazed and ungrazed sites required some careful planning. Differences could be attributed to grazing only if both sides of the fence had the same slope aspect and angle, and similar soils.

LOOKING ACROSS THE FENCES

The easiest and most legitimate comparisons were those where boundary fences of the sanctuary transected large and level benches and mesas. In 1990, we selected eight such places, scattered around the northern perimeter of the Ranch (Figs. 7 and 8). We then quantified grass canopy cover inside 20×50–centimeter quadrat frames, placed at two-meter intervals along tapes stretched perpendicular to boundary

FIGURE 7. Fenceline, looking south along eastern boundary of Research Ranch, 1997. (Photo by the authors)

FIGURE 8. Fenceline at northeastern corner of Research Ranch, 1997. (Photo by the authors)

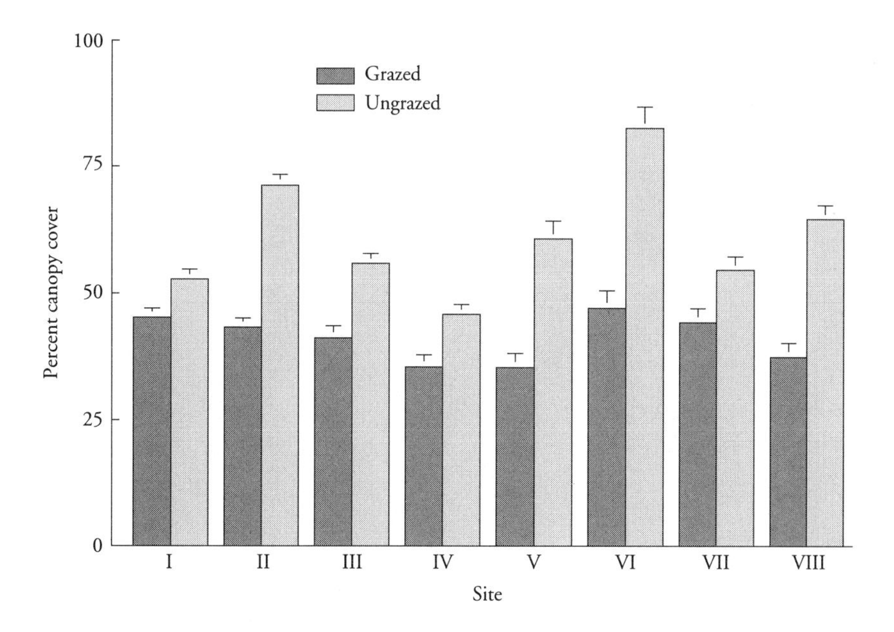

FIGURE 9. Grass canopy cover on grazed versus ungrazed quadrats at eight cross-fence sites around the perimeter of the Research Ranch in 1990, twenty-two years after livestock removal from the sanctuary. Bars are means with standard errors. (Redrawn from C. E. Bock and J. H. Bock 1993)

fences and running equal distances out onto grazed and ungrazed pastures. Our sampling effort included between 100 and 300 quadrats per site, depending on the size of the bench or mesa. Across all sites combined we measured plant canopy cover inside 650 grazed and 650 ungrazed quadrats.

Total grass canopy cover was significantly higher on the ungrazed side of each of the eight benches and mesas, although the magnitude of this difference varied substantially from site to site (Fig. 9). No cattle happened to be present on any of the eight grazed areas when we sampled them in August and September of 1990, so these differences cannot be attributed to recent clipping of the grasses by livestock. Rather, the data appear to reflect an overall long-term increase in grass cover resulting from over twenty years of livestock exclusion.

A Landsat satellite image of the Sonoita Valley taken in 1985 clearly shows the outline of the sanctuary, contrasted with adjacent grazed

FIGURE 10. Landsat satellite image of the Sonoita Valley and vicinity in 1985, clearly showing the darker outline of two ungrazed areas: the Research Ranch in the center, and Fort Huachuca to the right (east). (Image provided by Robert C. Marsett)

lands (Fig. 10). This image strongly supports our conclusion from work on the ground that grass cover is greater on than off the sanctuary. Darker colors indicate areas where vegetation cover is absorbing more light than would bare ground. Note the light color indicating a high level of reflection off the sandy creekbed in O'Donnell Canyon in the northeastern part of the sanctuary, and the generally lighter color, likewise indicating greater reflectance, of the adjacent ranches. Nearby Fort Huachuca, which also has been ungrazed for many years, is visible to the right of the sanctuary in the image.

Perhaps more interesting than overall cross-fence differences in total grass cover were specific differences for the ten most common grass

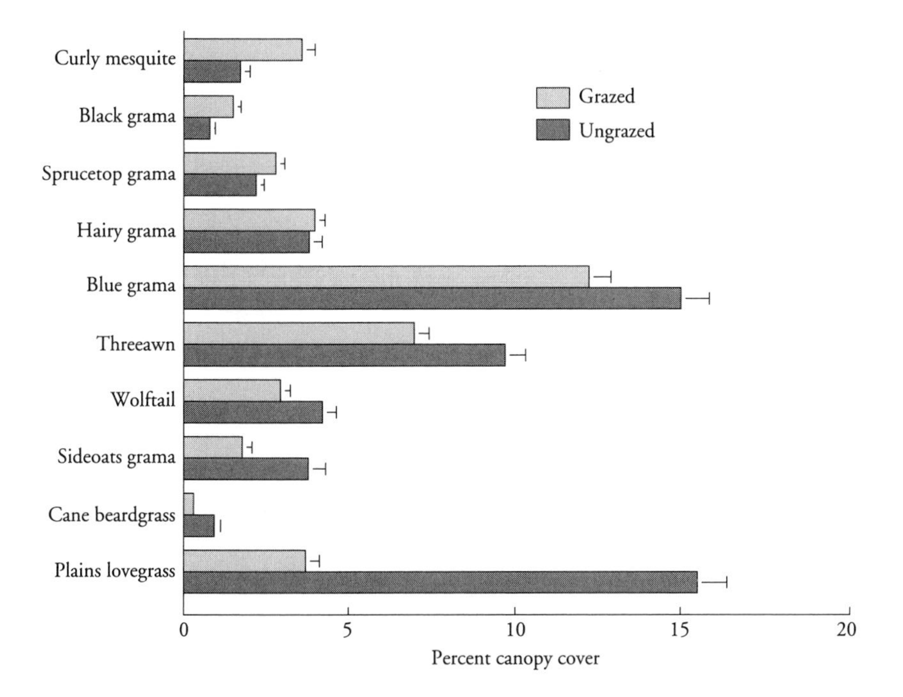

FIGURE 11. Percent canopy cover of ten grass species in 650 grazed versus 650 ungrazed quadrats on and adjacent to the Research Ranch in 1990. Grasses are ranked (from top to bottom) by increasingly positive responsiveness to absence of grazing. Bars are means with standard errors. (Redrawn from C. E. Bock and J. H. Bock 1993)

species individually (Fig. 11). We ranked the species in order of their responsiveness to protection from grazing, with plains lovegrass showing the strongest positive response and curly mesquite showing a strong negative response (i.e., it was more common in grazed areas). As the graph shows, six of the ten species responded positively to livestock exclusion; one, hairy grama, showed no difference; and three—black grama and sprucetop grama in addition to curly mesquite—were more common in grazed areas.

What factor or factors might explain why some grasses appeared to be favored by grazing, while others did better in the absence of cattle? Perennial grasses, such as those that dominate the Research Ranch, vary widely in their susceptibility to grazing owing to a variety of attributes. The most important of these is growth form. Let us, then, pause briefly for a lesson in grass morphology.

Bunchgrasses are deep-rooted species whose lateral growth is restricted to the formation of upright stems, called tillers, near the root crown. Older bunchgrasses will develop a large number of tillers, but the plant always will appear as a clump, usually surrounded by bare ground. By contrast, sod-grasses spread laterally by the formation of horizontal stems that grow in the soil (rhizomes) or on the soil surface (stolons). As these species grow, they can cover the ground and enmesh the soil with a continuous mat of vegetation, called a sod.

As a group, sod-grasses are more tolerant of grazing than bunchgrasses, since they have more growing points from which to regrow following a grazing event. Sod-bound soils are also much less vulnerable to erosion caused by water runoff after grazing. All grasses will lose the capacity for sexual reproduction if their flowering stems are chewed off, but this is of lesser consequence to a species that also can spread asexually by sending out rhizomes or stolons.

Stature, regardless of growth form, also is related to grazing vulnerability. Shorter species generally survive better in the presence of large herbivores because they lose a smaller percentage of their aboveground biomass to the grazing animal when they are clipped off. Furthermore, grazing may reduce cover of taller bunchgrasses that otherwise could outcompete the shorter species for light.

Taller bunchgrasses predominated historically in those parts of the American West where bison were scarce or absent, such as California and the Great Basin. These grasslands have been dramatically changed by cattle, horses, and sheep, and by the introduction and spread of exotic (nonnative) grasses that are more tolerant of grazing. Sod-grasses and shorter bunchgrasses predominated in the western Great Plains, where bison once were abundant and where historic livestock grazing has changed things relatively little.

The Sonoita Plain today supports a mixture of sod-grasses and bunchgrasses of varying heights. Our cross-fence data from the Research Ranch show that these grasses have responded to grazing as would be predicted based on their growth form and height (see Fig. 11). Only two of the ten most common species are sod-grasses

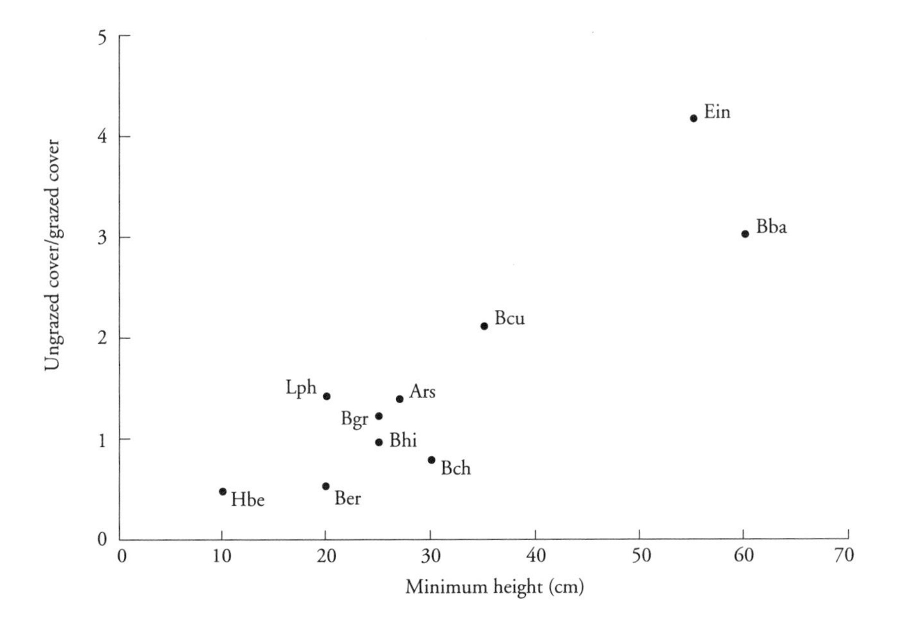

FIGURE 12. Ratio of mean canopy cover on ungrazed quadrats divided by mean canopy cover on all grazed quadrats, plotted against minimum height at flowering, for ten grass taxa on and adjacent to the Research Ranch in 1990. Acronyms are the first letter of the genus and first two letters of the species name (see Table 1), except for threeawns (Ars). (Redrawn from C. E. Bock and J. H. Bock 1993)

(curly mesquite and black grama), and they have been the most positively affected by livestock grazing. Furthermore, among these and the eight common bunchgrasses there was a striking positive correlation between height at flowering and response to livestock exclusion (Fig. 12). The three tallest bunchgrasses—plains lovegrass, cane beardgrass, and sideoats grama—apparently have been the most negatively affected by livestock, as would be predicted based on their height and growth form.

BLUE GRAMA

One of the most interesting grasses on the Sonoita Plain is blue grama. This widespread grass is the dominant species across most of the western Great Plains of North America. Blue grama also is very resistant to the effects of grazing. While not a true sod-grass, it is short-statured and, under grazing pressure, grows laterally to form a sodlike mat. Blue grama is today the dominant grass on the Sonoita Plain as a whole, and it remains about as common on the sanctuary as off (see Fig. 11).

The abundance of blue grama in the Sonoita Valley could be taken as evidence that the Arizona high plains evolved with large grazing mammals and that the impacts of livestock have therefore been minimal. Paleontological data, as well as our findings from the Research Ranch, suggest otherwise. First, there is no evidence that bison were present in any significant numbers in southeastern Arizona, at least for the past ten thousand years. Other ungulates such as pronghorn and deer were and are present, but they are not primarily grass feeders. Second, while blue grama remains common on the sanctuary, it is not among the species that have increased most since livestock removal.

Growth of taller bunchgrasses appears not to have resulted in any decline of blue grama on the Ranch, so far. However, such a decline may well occur if species such as plains lovegrass, cane beardgrass, sideoats grama, and wolftail continue to increase under protection from domestic grazers. This would make blue grama a transitory, easily displaced species that gives way to other native grasses in time, rather than the "ice cream plant" that range managers value for its persistence and palatability.

A central goal of our work at the Research Ranch has been reconstruction of the prehistoric Sonoita Plain, if not in fact then at least in our mind's eye. The case of blue grama illustrates the challenge, the risk, and the potential value of such an exercise. Assume that the trajectory of grassland changes seen today on the sanctuary will continue into the future. Assume that eventually blue grama will begin to be crowded out by the taller bunchgrasses, as seems very likely. Assume that this same trajectory can be taken as a view backward as well as forward in time. Certainly these are bold assumptions; yet they lead us to a tantalizing conclusion, and one that we cannot resist making. It seems likely that the very characterization of the Sonoita Valley as a blue grama grassland is an historical artifact—a consequence of centuries of grazing by livestock.

Grasslands of the Sonoita Valley and vicinity in southeastern Arizona have been variously described as desert grassland, semidesert grassland, high desert sod-grassland, or plains grassland, the latter especially because of the local abundance of blue grama. Based on our work at the Research Ranch, we believe there is developing here a landscape dominated by taller bunchgrasses that has not been recognized as a distinct plant community type in the Southwest owing to the chronic and ubiquitous influence of livestock grazing. We expect further declines of shorter-stature grasses typical of desert, semidesert, and plains grasslands, especially black and blue grama, as taller bunchgrasses continue to become more abundant on the sanctuary. In addition to plains lovegrass, sideoats grama, and cane beardgrass, mid-height bunchgrasses increasing on the exclosure include tanglehead (*Heteropogon contortus*), crinkleawn (*Trachypogon secundus*), Texas beardgrass (*Schizachyrium cirratus*), and woolly bunchgrass (*Elyonurus barbiculmus*).

We tentatively name this new community the *Madrean Mixed-Grass Prairie*, in light of its predominant growth form, its affinities to the flora of the Sierra Madre Occidental of northern Mexico, and its elevational position just below the oak-dominated Madrean Evergreen Woodland of the southwestern sky islands.

The reader need not accept this scenario to appreciate the value of livestock exclosures such as the Research Ranch. What the evidence shows with certainty is that the sanctuary today is beginning to support distinctive grasslands, dominated by taller bunchgrasses that are not as well represented on adjacent operating cattle ranches. The differences are clear from the air (see Fig. 10). These grasslands in turn support dramatically different assemblages of animals. The sanctuary is becoming just that—a place where plants and animals that are relatively intolerant of the activities of domestic grazers can find refuge and thrive.

A PROSPECTUS

The present polarization between ranchers and environmentalists is as understandable as it is unfortunate. Environmentalists rightly point to the historically destructive activities of livestock. They are justifiably angered by the apparent arrogance and aggressiveness of some ranchers, who regard their rights to public rangelands as paramount. Environmentalists note the dubious economic significance of grazing on western rangelands, and they probably are right in claiming that cessation of all grazing in the region would have a minimal impact on the nation's supply of red meat.

From the other perspective, ranchers are understandably offended by the activities of environmental groups and individuals perceived as interfering outsiders, ignorant and unconcerned about ranching families and their livelihoods. Ranchers point out that their economic future depends on proper care of the land. They are proud of their cultural roots, moreover, and many have an appreciation for the land that goes well beyond economics.

We are naive neither to the social dynamic as it presently exists between the conservation and livestock growing communities, nor to the real and substantive matters at stake on both sides. As environmental biologists, our greatest concern is for protection of grasslands and other western ecosystems. Nevertheless, we issue a plea for some sort of compromise in the new war for the West. There is common ground. We have found many ranchers to be keenly aware of, and interested in, the natural world they occupy. Major threats to western American landscapes include suburbanization, energy extraction, old-growth logging, and other unsustainable land uses. Ranchers, if they mean it about wanting to maintain their lifestyles and culture, are as threatened by these forces as is the natural world.

To us, the issue of livestock grazing in the American West has become unnecessarily burdened with dogma and accompanying rhetoric. A fair amount of this is attributable to the lobbying groups on both sides whose livelihood depends on perpetuating the controversy. We offer the following generalizations, in the hope that their acceptance by all parties might hasten development of a rational policy regarding grazing on western public rangelands and a truce in the war:

1. Historical grazing on fragile grasslands, especially those in very arid parts of the Southwest, has resulted in their virtual

destruction and desertification. It is unlikely that they can be returned to their prehistoric condition even in the absence of livestock without implementation of costly and innovative restoration techniques.

2. Among the grasslands that remain, those having a prehistoric association with bison, the continent's predominant native grazer, are relatively tolerant of the effects of livestock. These include primarily those grasslands of the central and western Great Plains. Rangelands in most of the Intermountain West, the Southwest, and the Pacific Coast existed since at least the Pleistocene without large herds of native grazers present, and so they are relatively intolerant of the activities of domestic livestock. Grazing policies must be tailored to reflect this fundamental difference.

3. Livestock are neither universally beneficial nor absolutely harmful to vegetation and wildlife. Rather, there exist in almost every region of the West groups of plants and animals that thrive in the presence of large grazing mammals, and other groups of plants and animals that clearly do not. Livestock usually function in the role of a keystone species; that is, they have a controlling influence on the structure and function of those grasslands where grazing occurs. By their activities, livestock determine which species will become common and which will become rare.

4. Given the virtual ubiquity of livestock over much of the landscape of the West, those species of plants and animals relatively intolerant of the activities of large hooved (keystone) ungulates have fewer places left to live than their relatives that thrive in the presence of grazing. Therefore, creation and maintenance of large permanent livestock exclosures would do much to enhance the biodiversity of our public grasslands.

5. There is no economic or ecological justification for predator control on western rangelands, nor for opposition to reintroductions of predators such as wolves. There is no single

issue where public opinion is more at odds with the livestock industry, and for good reason. It seems to us that predator control has much less to do with livestock losses than it does with holding some sort of relict dominion over the land. Those days are gone in the West, and they should be.

6. There is no need for absolute exclusion of livestock from public lands. Although certain particularly fragile areas, especially wetlands and riparian habitats, deserve this sort of protection, elsewhere a heterogeneous mosaic of grazed and ungrazed landscape units would provide for the greatest abundance and variety of native species, and this level of grazing should be sustainable over the long term.

As active environmentalists, we have and probably should have a radical side. There is something clean and alluring about the philosophy of the pure environmental anarchist. As Hayduke might well have said, "Let's just get these cowboys and their four-legged vermin off our land, and let it heal." Sense and sensibility tell us that this worldview might make us feel good, but actually accomplishing something will require more humanity and compromise. We have found ranchers generally more interesting than most lawyers, lobbyists, or legislators. It is with them that we choose to work toward a goal of conserving what is left of the natural landscape of the American West.

Five

HERPS

There are about eleven thousand species of reptiles and amphibians in the world. Collectively they are called herps, and the people who study them are herpetologists. Herps constitute less than 1 percent of the animal species known to exist. However, they have received attention all out of proportion to their numbers because herpetologists are relatively abundant among biologists and many of them are wildly enthusiastic field naturalists. The result is that we know a great deal more about the natural history of reptiles and amphibians than we do about most other kinds of living things, with the exception of birds and mammals.

This knowledge base makes reptiles and amphibians potentially powerful indicator species. A good indicator species is one whose ecological requirements are precise and very well known, so that data on its relative abundance can be used as a surrogate to other environmental measurements that are more costly and more difficult to obtain. There is no point in choosing an indicator species whose presence tells something about the environment we can readily see for ourselves. We do not need to count alligators to determine whether we are in a swamp. On the other hand, we might count sea turtles as an indicator of the overall health of certain tropical marine ecosystems.

Indicator species, properly chosen and studied, may provide clues about past and future as well as current environmental conditions. Sanctuaries and preserves such as the Research Ranch are particularly important in this regard, if they represent remnant patches of ecosystems that used to be more common. Perhaps only in such places will a putative indicator species reveal things about its true environmental roots and requirements. Three examples from the herpetofauna of the Research Ranch illustrate this point.

THE BUNCHGRASS LIZARD

In making our comparisons between grazed and ungrazed habitats on the Sonoita Plain, we attempted to choose broad expanses of relatively uniform landscapes, transected by boundary fences between the Research Ranch and adjacent properties. Only in such places could we be reasonably certain that ecological differences across the fences were entirely attributable to the presence versus absence of livestock, and not to other factors such as different soils or topography. Two of the places where we did the most work were the North Mesa and East Mesa, in the northeastern portion of the sanctuary (see Map 2). We compared abundances of grasses, shrubs, herbs, rodents, grasshoppers, and birds across boundary fences here. We also studied the effects of wildfire and the planting of nonnative (exotic) grasses in these areas.

During our many hours of fieldwork on the North and East Mesas between 1981 and 1987 we had numerous fleeting glimpses of small and superficially nondescript lizards scuttling off through the ungrazed grasslands. The lizards were a topic of casual conversation, but we made no attempt to capture one or to determine how many or what particular species might be there. In retrospect, it was an odd thing not to have done, except that lots of other things were keeping us busy, and neither we nor any of our field assistants happened to be passionate herpetologists.

In 1988 we finally captured several of these lizards. The animals we caught were about two inches long, not counting their tails. They were streaked and spotted in shades of gray and brown above; the females were pale below, and the males had brightly colored longitudinal bands

FIGURE 13. Male (left) and female (right) bunchgrass lizards. (Photo by the authors)

of salmon orange and vivid blue along their flanks and bellies. Based on these characteristics, we tentatively identified these animals as *Sceloporus slevini,* the bunchgrass lizard (Fig. 13).

Now, the bunchgrass lizard certainly is a grassland specialist, but it has been found north of Mexico almost exclusively at high elevations (up to 10,000 feet) in isolated montane meadows of the sky islands of southeastern Arizona and southwestern New Mexico. There were a handful of records from lower elevations, including one from the Sonoita Valley. However, these sightings were very rare, and herpetologists routinely have described the bunchgrass lizard as a species of the mountains. Yet here it was, apparently the most abundant reptile in desert grasslands of the Research Ranch. Had we misidentified our lizards, or had we come upon something new? It was time to get some expert help.

Hobart Smith, a colleague of ours at the University of Colorado, has the distinction of having published more books and papers about rep-

FIGURE 14. Hobart Smith, 1999. (Photo by the authors)

tiles and amphibians than any other living herpetologist (Fig. 14). At last count, his resume included *well* in excess of a thousand titles—and counting. Dr. Smith also happens to be a particular expert on lizards of the genus *Sceloporus.*

Hobart Smith visited the Research Ranch in August 1989, and we worked with him in the field, doing a thorough and systematic count of lizards on the North and East Mesas, on both sides of the sanctuary's boundary fence. The only common lizards on these grassland benches were indeed *Sceloporus slevini.* Furthermore, of forty-four individuals captured and so identified, only three were on the grazed side of the fence.

Here, then, was something new. One of the commonest reptiles on the Research Ranch was a species supposedly abundant only in neighboring high-mountain meadows. The implications were clear. If grasslands on the sanctuary today represent something like conditions in the Sonoita Valley as a whole two centuries ago, then the apparent restriction of the bunchgrass lizard to the mountains probably is a historical artifact of habitat modification by livestock in the lowlands. The present-day concentration of bunchgrass lizards in high mountains probably is an ecological indicator of just how much grasslands in the Sonoita Valley have changed since the introduction of cattle and horses.

Why might the bunchgrass lizard require heavier grass cover than that normally provided by grazed sites? There are a number of possibilities, but we have circumstantial evidence that these animals may use the dense grass as a refuge from predators. A few miles west of Elgin, Smith explored the abandoned and scattered remains of an old ranch building and corral. The surrounding land was heavily grazed and largely barren of bunchgrass lizards. However, they were common in the vicinity of the old buildings, where they took refuge under the numerous boards and posts lying about. Apparently these lizards had no trouble finding food or other life requirements in grazed habitat, as long as they had a place to hide.

There is one other intriguing aspect to the case of the bunchgrass lizard. Species in the genus *Sceloporus* are highly variable in terms of their reproductive biology. Some breed only once per year, while others breed multiple times in a single season. Some are long-lived and slow to mature, while others mature quickly but are short-lived. Some species lay eggs (oviparity), while others give birth to live young (viviparity). Herpetologists have attempted to subdivide the genus into groups with similar life-history traits, and then to correlate those traits with environmental conditions.

Sceloporus slevini apparently belongs to a subgroup of species that share a suite of life-history traits. Most members of this subgroup reach sexual maturity in one or two years, live in relatively moist high-elevation habitats, and are viviparous. Viviparity in particular seems to be a trait of high-elevation *Sceloporus*, probably because of the advantages of retaining the developing embryo in the body of the female in cold, oxygen-scarce environments.

The bunchgrass lizard fits some of the pattern but not all of it. Specifically, while they do mature in one year and do occupy high elevations, they are egg-layers. A possible explanation for this anomaly is that *Sceloporus slevini* may have been primarily a species of lowlands prior to historical habitat changes brought by the introduction of livestock.

RATTLESNAKES

"Mommy, there's a little rattlesnake down
in a bunny rabbit hole. Come see!"

LAURA KAY BOCK, AGE 3, 1974

Next to rain, the most common subject for general conversation in the Sonoita Valley is snakes. Bringing up snakes in that part of the world is like bringing up bears at a meeting of National Park Service rangers. Everybody has a story.

There are three common species of vipers on the Research Ranch, the black-tailed rattlesnake (*Crotalus molossus*), the western diamondback rattlesnake (*C. atrox*), and the Mojave rattlesnake (*C. scutulatus*). We have made no systematic survey of their distributions, abundances, or preferred habitats on the sanctuary. However, an encounter with a rattlesnake is a memorable event, usually shared with others afterward, and over the years some clear patterns have emerged.

The black-tailed rattlesnake is distinguished from the other two species by the absence of white rings on its all-black tail. We have usually encountered them on the southern part of the sanctuary, where they are associated almost exclusively with rocky outcrops and ledges on oak slopes or along streams. They are comparatively docile animals—unaggressive, retiring, and not inclined to rattle.

The diamondback is the largest rattlesnake in western North America, occasionally reaching lengths over 7 feet (2 m). The tail is banded in black and white rings of about equal width. At least on the Ranch, most diamondbacks we've seen have had a grayish or dusty color. These large animals are widespread on the sanctuary, but we have encountered them most frequently along washes and in floodplains. Diamondbacks have a reputation for being particularly aggressive, rattling furiously and defending their ground when encountered.

The Mojave rattlesnake also has a banded tail, but the black rings are narrower than the white ones. Individuals encountered on the Research Ranch have usually had an overall background color of greenish yellow. This is the species of rattlesnake we encounter most frequently on the sanctuary. It is most abundant in upland grasslands and oak savannas. Mojave rattlesnakes usually rattle when encountered, and they quickly assume coiled defensive postures. However, they appear very reluctant to strike, which is fortunate, because this species has an unusually high concentration of deadly neurotoxin in its venom.

At one time, the largely complementary distributions of the three rattlesnake species on the Research Ranch would have been interpreted by ecologists as evidence that they were competing with one another for limited resources. Presumably each species would be most abundant in the habitat where it was the most effective predator, and where it could outcompete the other two for prey.

Today we have learned to be more cautious about attributing complementary distributions to interspecific competition. Often such patterns are simply expressions of the independent habitat preferences of particular species. We can hypothesize a competitive relationship among these snakes, but we could test this hypothesis only by performing a field experiment. For example, we could capture and remove Mojave rattlesnakes from an upland grassland, and then determine if either black-tailed rattlesnakes or western diamondbacks moved in to replace them.

Intrusive manipulative experiments such as removing snakes from large areas are the sorts of things that should not happen on sanctuaries designated as ecological control areas. The policy at the Ranch has always been, and should always be, to minimize human impacts and to make fieldwork as purely observational as possible. Significant alteration of the ecosystems represented would inevitably compromise their value as natural benchmarks.

Whether they are competing or not, rattlesnakes on the Research Ranch seem to have clear habitat preferences, and they probably can be taken as indicators of the quality and availability of those habitats. Important fieldwork elsewhere in the Southwest supports this conclusion.

The country east of the Chiricahua Mountains in extreme southeastern Arizona and southwestern New Mexico is a mixture of semidesert grassland and Chihuahuan desert scrub. Within the past thirty years this area has experienced considerable desertification as scrub vegetation has replaced former arid grasslands. A snake survey undertaken here between 1959 and 1961 revealed that Mojave rattlesnakes outnumbered diamondbacks by about three to one. By 1989, the two species had become about equally abundant, and diamondbacks had largely replaced Mojave rattlesnakes in areas that had shifted from grassland to Chihuahuan desert. These results support the conclusion that Mojave rattlesnakes are indicators of intact desert grassland habitat and that their decline is a sign of desertification. Continuing abundance of the Mojave rattlesnake on the Research Ranch, therefore, is another indicator of the quality of its grasslands.

THE LEOPARD FROG

At the southern end of the North Mesa is a small spring. It is nothing more than a slow but steady flow of clear water that trickles out of the ground at the base of a small rock ledge. In May 1921, James L. Finley established a homestead here, and he dug a small pond to collect water from the spring. That pond is called Finley Tank, and it remains one of the few permanent water sources on the Research Ranch (see Map 2).

Finley Tank is a true oasis, surrounded by a small grove of willows and cottonwoods. Especially in the hot dry days of May and June, it stands out as a patch of emerald green against a background of browns and grays. Coyotes, deer, and javelinas come to Finley Tank in the evening to drink. There usually is a porcupine living in one of the willow trees, and it is not uncommon for a great horned owl to roost there during the day.

The pond itself is an aquatic island in a sea of desert grass. It is part of a very small drainage system that is contained entirely within the sanctuary boundaries. We were surprised, therefore, to discover the variety of freshwater organisms living in the pond, despite its isolation. Perhaps James Finley himself stocked his pond with fish. Perhaps birds accidentally brought in seeds and the eggs of aquatic insects.

When we first visited Finley Tank in March 1974, it was cool and clear and full of Chiricahua leopard frogs and their tadpoles. Ten years later it was just as cool and clear, but the leopard frogs were gone. Very small, isolated populations of anything have a reasonably high risk of going extinct, just by chance, and so we were disappointed but not overly surprised when the leopard frogs vanished from Finley Tank. Then we began hearing stories about mysterious declines and disappearances of leopard frogs and other amphibians all over.

From alpine toads in Colorado to salamanders in Costa Rica, amphibians seemed suddenly to be in trouble in lots of places. Herpetologists and environmental biologists quickly began searching for causes. Sometimes the declines could be attributed to, or at least correlated with, local disturbances such as water pollution or habitat destruction. However, these events were so widespread geographically, and involved such a variety of species, that attention soon turned to a search for factors operating at a continental or perhaps even a global scale.

As we write this, herpetologists have come to no consensus regarding the continuing declines of amphibians. Some theory and evidence suggest causes such as acid rain, increased ultraviolet radiation, or atmospheric global circulation of toxic chemicals. Amphibians are thought to be particularly vulnerable to these sorts of factors because of their rather delicate, moist skin and because in most cases they must return to freshwater to breed.

Environmental scientists are themselves vulnerable these days, and most are understandably reluctant to announce that the sky is falling if they are not absolutely sure that it is. But we would be fools not to take the amphibian disappearances seriously, as possible early indicators of widespread environmental degradation that could affect us all.

It is difficult to imagine a population more isolated against intrusions from the outside world, and therefore living under more favorable circumstances, than were the late leopard frogs of Finley Tank.

Six

ISLANDS OF FIRE

TO BURN OR NOT TO BURN

IN THE FLINT HILLS OF eastern Kansas there is a grassland preserve called the Konza Prairie. Owned by the Nature Conservancy and managed by Kansas State University, this site is one of the largest remaining tracts of true (or tallgrass) prairie—that treeless, grasses-up-to-your-nose landscape that originally characterized the eastern third of America's Great Plains. It once extended from southern Manitoba to northeast Texas, but most of the tallgrass prairie was lost to the steel plow during the nineteenth century because of the extraordinary agricultural potential of prairie soils and climate. The Flint Hills are too rocky to plow, and so they and a few other places remain to give us a glimpse into the dynamics of this once great biotic province.

The Konza Prairie is remarkably similar to the Research Ranch in terms of its size, topographic configuration, and present purposes. Both areas are managed as natural laboratories and biodiversity preserves, where the structure and function of grassland ecosystems can be conserved and studied with a minimum of human interference. Both are islands in an agricultural sea, surrounded and isolated by much larger blocks of land in distinctively different ecological condition. Both are limited in what they can tell us about the prehistoric grasslands they are presumed to represent, because of their present isolation.

In other ways Konza and the Ranch are poles apart. We know that periodic droughts, grazing by hooved mammals, and fire are the three driving forces that determine the distribution, composition, and function of most of the world's grasslands. Yet not all grasslands are equally influenced by these three forces, and in many ways the tallgrass prairies of the American heartland and the desert grasslands of the Southwest are at opposite ends of an environmental spectrum. Most importantly, rainfall in the Sonoita Valley averages less than half that in eastern Kansas. A second distinction is that bison and other native hooved grazers have been scarce or absent from the desert grass over the past ten thousand years, whereas they were present and probably a major ecological force in the original tallgrass. Bison have been reintroduced at Konza to examine this possibility.

Another important difference between the Flint Hills and the Sonoita Plain is in the effects of fire. The Konza preserve is divided into a series of watersheds separated by roads that serve as firebreaks, and nearly all of these watersheds are deliberately burned on a schedule ranging from one to twenty years. The objective is to understand the ecological consequences of burning tallgrass prairie at different seasons and with different frequencies. At a more fundamental level, however, the importance of burning the Konza Prairie has been absolutely clear from the beginning.

Up and down the original extent of tallgrass, average annual rainfall is more than sufficient to grow trees. In the absence of fire, most tallgrass prairies simply disappear and become woodlands. An abandoned cornfield in western Iowa or eastern Kansas or central Oklahoma does not revert to prairie. Rather, it becomes what the farmer will call a wood lot—a combination of ash and oak and elm, with blue jays and cardinals and red-headed woodpeckers: that is, everybody's Middle American Forest.

Does or did fire play any sort of equivalent role on the plains of the Sonoita Valley? Lightning and human-caused wildfires have been relatively frequent events at the Research Ranch, but there are large portions of the sanctuary that have not burned since the cows were taken off in 1968, and probably not for many years before then. These unburned areas have not become forests or woodlands, as they long since would have done in eastern Kansas. In fact, a first-time visitor to the Research Ranch would be hard-pressed to distinguish them from sections of the sanctuary that have burned in recent years. While fire in the Sonoita Valley clearly does not play the critical role it does in eastern Kansas, nevertheless we have found that it has important effects that may influence the long-term viability of its grasslands.

In any natural ecosystem, two ingredients are necessary for fire: fuels dry enough to burn and a source of ignition. Fires are especially likely to occur in southeastern Arizona from June to early July, when temperatures are high and humidity low, when the grass is still brown, and when dry lightning comes at the beginning of the summer monsoon.

Fires have been scarce and small in recent times in the Sonoita Valley, for three reasons. First, people attempt to control them out of concern for their property. Second, the landscape is fragmented by firebreaks such as roads, so when fires do start they usually are quickly contained. Finally, and most critically, in most places livestock grazing has removed what otherwise might burn. Domestic grazers have pretty much fire-proofed the Sonoita Valley by eating the fuel.

The question of fire is far more than academic for managers of the Research Ranch; in fact, it presents a significant management dilemma. Recall that the overall strategy at the sanctuary is to provide opportunities for natural ecological patterns and processes to express themselves to the extent possible in the absence of human intrusion. But what does this mean in terms of fire? Certainly a natural lightning-caused wildfire should be allowed to burn. What about an accidental human-caused fire? There is substantial evidence that Native Americans set fires in the Southwest both deliberately and accidentally, so anthropogenic burning also has been a part of the regional environment for thousands of years.

Wildfires have occurred but have not been permitted to burn completely free on the Research Ranch. This is because the sanctuary is within a region where personnel from the Coronado National Forest and the local volunteer fire department attempt to control all fires before they spread off the sanctuary and threaten our neighbors' homes and the forage for their cattle.

In our years at the Ranch fires rarely have ignited on adjacent properties and then spread onto the sanctuary, because of reduced fuels on grazed lands. Therefore, we can be virtually certain that grasslands on the Research Ranch are burning today less frequently than they did prehistorically. This raises the possibility that the sanctuary should be burned deliberately, as is done at Konza, to simulate the physical environmental arena in which the flora and fauna of the Sonoita Plain evolved and once thrived.

What has happened when, patchily and periodically, parts of the Ranch have risen from their own ashes? One of our central research objectives has been to measure the ecological effects of wildfires that have occurred here since 1968, with the goal of providing information useful to managers who must decide whether to burn or not to burn, and when, and how often.

SACATON

There is an intriguingly persistent story that Coronado rode through grass as high as the back of his horse when he traveled north out of Old Mexico in 1540 and perhaps passed within view of what is now the Research Ranch. Upon first consideration, this notion seems as fanciful as the seven golden cities of Cíbola that he failed to find. Arid southwestern plains simply could not grow grass that high, even in some long-forgotten prehistoric condition.

However, there is one true tallgrass in the Southwest, called big sacaton. It grows over 6 feet (2 m) high, and often in nearly pure stands. Sacaton thrives only in broad floodplains, where water meanders gently enough not to erode away the deep soil sediments it requires, but often enough to keep it alive and to allow it to grow so tall. Because Coronado followed the watercourses, he almost certainly rode in sacaton at least a part of the way, and so the story may well be true.

Few habitats are more endangered in southeastern Arizona than sacaton bottomlands, primarily because flash floods have torn away the grass clumps and carried off the soil. The water then becomes channelized into rocky stream- and riverbeds, usually lined with riparian trees such as cottonwood, ash, willow, sycamore, and walnut. Intact sacaton stands probably burned often and with great intensity, killing trees that happened to have colonized the watercourses. Certain southwestern riparian woodlands, rich and valuable as they are for birds and other wildlife, may be a historic artifact of stream channelization, the resulting loss of sacaton, and consequent reductions in fire frequency and intensity.

Gale Monson, a longtime student of Arizona birds and their environments, once told us that the Research Ranch supported the best remaining stands of sacaton grass he had ever seen (Fig. 15). Of all parts of the sanctuary that have burned since 1968, none have done so with more dramatic ecological effect than this particular habitat.

On a hot and clear day in June 1975, residents of Elgin looking to the south saw a low-flying aircraft jettison a large quantity of some unknown liquid, bank steeply upward, and fly away. Minutes later smoke was seen to rise, and a fire of increasing intensity began to move down O'Donnell Canyon. It burned into sacaton grasslands that are abundant at the confluence of the O'Donnell, Post Canyon, and Turkey Creek drainages. The U.S. Forest Service responded and stopped the fire that afternoon, before it reached any buildings on the sanctuary.

Sacaton fires are spectacular. The burning grass clumps almost explode, shooting flames over twenty feet into the air. All aboveground vegetation is completely combusted. What came to be known as the "Bomber Burn" consumed nearly nine hundred acres of prime sacaton grassland, killed the largest cottonwood trees on the sanctuary, and left behind a totally blackened landscape. In 1975 and 1976 we compared this burn with adjacent sacaton stands spared by the fire. Our goal was to determine the magnitude and persistence of the impacts of fire in this unique habitat.

Unburned mature sacaton is a rather simple community. The one tall grass dominates the vegetation. Cotton rats are the only common rodents, and they make characteristic ground-level runways through sacaton and feed on its foliage. A host of predators key on the cotton rats, including coyotes and western diamondback rattlesnakes. A diet of cotton rats apparently grows big snakes. We had some adventures

FIGURE 15. Jane Bock in an O'Donnell Canyon sacaton stand, 1978. (Photo by C. Bock)

working in sacaton, because the snakes were attracted to our rodent live-traps and because we often failed to notice each other soon enough for evasive action in the tall grass.

Other animals characteristic of unburned sacaton included species with various requirements for heavy cover, such as the Montezuma quail, common yellowthroat, blue grosbeak, Botteri's sparrow, and javelina.

Fire radically but temporarily changed the sacaton community. It killed only a few grass clumps, but it retarded growth in all of them throughout the first postfire growing season. Released from the dominance of sacaton, other grasses and wildflowers, with their copious seed production, increased dramatically on the burn as soon as the monsoon rains began. Very different animals became abundant. Cotton rats virtually disappeared and were replaced by seed-eating species preferring areas with less cover—especially the Merriam's kangaroo rat and hispid pocket mouse. Avian predators such as the American kestrel and northern harrier hunted the burn more frequently than adjacent unburned areas, probably because their prey were more visible. A variety of different birds concentrated on the burn in the fall and winter of 1975–1976, particularly seed-eaters such as the mourning dove, eastern meadowlark, Savannah sparrow, vesper sparrow, Cassin's sparrow, and white-crowned sparrow.

Except for the loss of the cottonwoods and considerable damage to some velvet ashes in upper O'Donnell Canyon, effects of the Bomber Burn were as short-lived as they were dramatic. Most grasses, including sacaton, are not killed by fire because the tissues from which they resume growing are protected just below ground level. Fourteen months after the burn, sacaton had returned to its preburn dominance, and the pulse of postfire plants and animals had begun to wane.

Many ranchers and land managers recommend that sacaton be burned in late winter or early spring, to provide tender new growth for cattle forage and to prevent permanent damage that might result from fire in a hotter, drier season. Studies of the Bomber Burn and other sacaton fires on the Research Ranch suggest that burning at the natural fire time (in summer, just before the monsoon) may have advantages. An early-summer "hot" fire does kill some sacaton, but this opens the habitat and stimulates foliage and seed production by other native grasses and wildflowers, and these in turn provide food for a variety of insects, birds, and rodents. Within a very few growing seasons after a fire, sacaton will return to its former density unless other disturbances ensue. However, intensive grazing of sacaton can lead to separation and weakening of individual plants, making them vulnerable to loss through erosion.

From a conservation perspective, sacaton management might best include a mosaic of stands burned in different years, with perhaps an average four-year fire rotation. This would give ample ecological opportunity for those species dependent on mature stands, as well as for those keyed to circumstances in the first year or two after fire. Given the inherent combustibility of dry sacaton and the frequency of lightning just before the monsoon, this undoubtedly would have been the natural state of affairs prior to fire-prevention policies imposed over the past hundred years.

UPLANDS

July 16, 1987, started out clear and unseasonably windy. The monsoon was late that year, and conditions were very dry. In the afternoon a small thunderhead formed above Turkey Creek, near the southern boundary of the Research Ranch. No rain ever fell out of this particular cloud, but it did gather enough energy to discharge a bolt of lightning into the oaks below. A single tree exploded and burst into flame. Fanned by winds gusting to sixty miles per hour and fueled by the dry grass, the fire spread quickly north down Turkey Creek and up onto the East Mesa. Fire crews responded quickly and in force, but the intensity of the fire was such that it was not controlled until well into the morning of July 18, by which time it had blackened almost all of the eastern third of the sanctuary—nearly 2,500 acres.

The fire was easily stopped at the sanctuary boundaries, where it ran out of fuel. Interior fire lines were much harder to establish. Here, it took a combination of helicopters with water buckets, slurry bombers dropping their characteristic red plumes of fire retardant, and skilled teams of White River Apache firefighters on the ground (Fig. 16).

This grassland fire was like none we had ever seen. The leading firewalls ran at alarming speeds before the wind. Great vortices of flame and ash gathered in the lowlands and spun erratic paths up the slopes and onto the mesa tops, where they lost energy in the sparser grass. Flames from the sacaton lit up the night sky in shades of yellow and orange and red. We could have read a book outdoors at midnight, it was so bright, but nobody was doing any reading. We evacuated our house at East Corrals and moved ourselves, the dog, a pet tortoise, and that year's field data up to Headquarters. The fire burned to our very

FIGURE 16. Fighting the Big Fire, July 1987. (Photo by the authors)

doorstep, but thanks to the firefighters, including some high-ranking government officials working alongside local volunteers and our ranching neighbors, nothing was lost. It was an exciting time, and in the end the fire had very satisfying consequences.

By the summer of 1987 we already had conducted studies on many parts of the Research Ranch, and so it was inevitable that the Big Fire (as it came to be known) combusted a variety of former study plots. This created the circumstance for a definitive study of fire effects. Because an equal number of former plots escaped burning, we were in a position to examine results of an ideal, doubly controlled, field experiment. We had sufficient data to compare two sets of plots before either burned (the prefire control), and then to compare the same two sets after one burned (with the unburned set as a postburn control). Based on results of these comparisons, and studies of some earlier and smaller burns, we have learned quite a bit about how the flora and fauna respond to fire in upland grasslands in the Sonoita Valley.

The Big Fire, like most grassland burns, had predictable effects on the physical environment. By combusting standing dead plant material, it increased mineral content of the soil. While usable nitrogenous and carbon compounds are volatilized by fire, the remaining ash recirculates minerals that otherwise may be in short supply. This ash can act as a growth enhancer (fertilizer) for the postfire vegetation. Established herbaceous species are almost never killed in a grassland fire, because their growing points are close to or just beneath the soil surface, which offers them protection. Another encouragement to growth comes from the blackened soil surface, which increases early-season warming. These altered thermal and nutrient regimes can result in an especially productive grassland (particularly in terms of seed germination) in the first year or two after a fire.

Overall cover of perennial grasses declined for two growing seasons following the Big Fire, but returned to preburn levels during the third postfire monsoon. In general, threeawn and grama grasses were more negatively affected, while the more exclusively southwestern species, such as wolftail, plains lovegrass (see Chapter 11), and the bluestems, recovered more quickly. Cover and species richness of wildflowers increased on burned plots for two years, probably in response to the temporary reduction in competition from the grasses as well as the more favorable soil conditions.

There is a major debate among ecologists and paleobiologists about the role fire may once have played in keeping woody vegetation out of desert grasslands. Historical evidence suggests that native shrubs and small trees, especially mesquite, have increased dramatically in much of the lowland Southwest in the past one hundred years. Livestock grazing, absence of fire, rodent control, and climate change all have been implicated. It has been difficult to untangle the separate and interactive roles these forces may have played because of the absence of large control areas without livestock. We can scarcely pretend that the Research Ranch is such a place, given its own history as a cattle ranch. But by the summer of 1987 there had been nineteen monsoons without a cow present, and so it is possible the Big Fire burned with an intensity comparable to that of its ancient predecessors.

How did the woody plants respond? Only one common shrub, burroweed, was absolutely killed by the fire. However, it had become so common on unburned parts of the sanctuary that its loss equated to a major change in plant community structure. Even by the summer of 1995 burroweed had barely begun to recover from the Big Fire, so upland grasslands in the fire's path still retained a very different aspect than those elsewhere on the sanctuary. Other trees and shrubs on the Sonoita Plain have the capacity to regrow from ground level even if their tops are burned away, which means that fires have impacted them much more ephemerally. On the Research Ranch, the most important of these are mesquite, yerba de pasmo, and catclaw mimosa. For each, fire reduced the height and width of the plant crowns for two or three years but caused only limited mortality.

Overall, it appears that upland fire on a three- or four-year rotation would have the long-term effect of reducing woody vegetation in grasslands of the Sonoita Valley while sustaining or stimulating the abundance and variety of native perennial grasses and wildflowers. It is possible that repeated fire eventually would kill many woody plants. We suspect this was the situation before the time of Coronado.

As in the sacaton, upland fires on the Research Ranch sanctuary have favored some wildlife populations while temporarily reducing others. We found evidence of direct fire-caused mortality in only two groups: grasshoppers and rattlesnakes. Large mammals, birds, and rodents escape fire by fleeing or by taking refuge in their burrows. Songbirds increased more than tenfold in the first fall after the Big Fire, almost certainly in response to dramatically increased seed production by short-lived wildflowers and grasses. In contrast, densities of a variety of nesting sparrows declined for one or two years, until the grass cover they require had returned to its preburn abundance.

Any strategy for managing fire in desert grasslands generally, and on the Research Ranch in particular, should be one of balance. A mesquite tree is no less noble than a clump of plains lovegrass, nor is the loggerhead shrike nesting in its branches any less a part of the natural scene than a grasshopper sparrow nesting in the grass beneath it. There is no wisdom to a land management strategy that exclusively favors one native species over another. Burning all the sanctuary at once, or burning different parts of a mosaic in different years, but all of them often, would be very bad ideas. However, neither plan would be less desirable or less natural than permanently excluding fire from the whole place.

If we could travel back in time to the Sonoita Valley of the eighteenth century, we probably would find a mosaic of landscapes in various stages of postfire ecological succession. Fire would have spared some patches for decades, purely by chance, and here the woody plants and their associated fauna would be relatively abundant. Most patches would have burned more frequently, perhaps an average of once every three years. Here the native grasses would predominate. Returning this sort of fire disturbance pattern to the Research Ranch will be challenging, given soil and resulting grassland losses dating mostly from the late nineteenth century, but it also will provide maximum opportunity for conservation of the regional biota in its full variety and abundance.

THE BROPHY PRESCRIPTION

Land managers refer to deliberately set fires as "prescription" burns. A prescription is some combination of environmental conditions where fire is likely to have predictable and desirable consequences, and where it can be controlled. Prescriptions include such things as fuel moisture levels, humidity, temperature, wind speed, time of day, season of the year, and whether to run a fire with or against the wind. The purpose of a prescription burn is to return fire to ecosystems where it once was a natural event in order to facilitate desirable management consequences, such as keeping a grassland relatively free of invading trees and shrubs.

When the U.S. Forest Service, Bureau of Land Management, or another public agency puts on this sort of controlled burn, dozens of people get involved. They not only design the prescription and determine if conditions are right before ignition, but they also are on hand and properly equipped to make sure that the fire goes only where it is supposed to go and stops when it is supposed to stop. We have observed and studied prescription burns throughout much of the West, and we have been uniformly and increasingly impressed with the care and attention to detail that fire professionals bring to their task. Managers of the Research Ranch and other grasslands should not implement prescription burning without this sort of help.

There are, however, more casual approaches to burning in the Sonoita Valley. On a spring morning in 1986 we were giving a tour of the Ranch to Peter Berle, newly appointed president of the National Audubon Society. It was a day to make good impressions, not only as to the bounty of the place, but also about the quality of our stewardship. At one point during the tour we were driving a road just inside the northern boundary of the sanctuary, where it shares a fence with the adjacent San Ignacio del Babocomari cattle ranch. We drove east to the edge of the North Mesa, in order to give President Berle a look down onto the sacaton grasslands bordering O'Donnell Creek below (see Map 2).

To our complete surprise, the whole place was on fire. There had been no lightning, and there was nobody around. It was apparent that the fire had started on the Babocomari side of the fence, and that it was burning slowly but steadily south onto the sanctuary. If Peter Berle was dismayed about what was happening to some of his prime real estate, he was kind enough not to show it. We started telling him about the natural and beneficial effects of fire in the Sonoita Valley, and we hoped he wouldn't ask how this particular one got started—because we had a sneaking suspicion.

Sacaton provides nutritious forage for livestock, but cattle eat it with enthusiasm only when they can get at tender new shoots—such as start to grow after fire, and especially before the summer rains stimulate new growth in other grasses that they like better. Bill Brophy, of the Babocomari, liked to burn his sacaton in the spring so that it greened up in May and June when there was not much else around to eat.

We had, it seemed, come upon a Brophy "prescription" burn. As usual, the plan had been to pick a clear dry day, throw a cigarette into a big clump of grass, make sure it was going good, and then head home for a cup of coffee.

When we got back to Ranch headquarters there was a phone message from Bill. Darned if the wind hadn't shifted, and it looked like his burn had picked a direction he hadn't planned on. We might want to call the local volunteer fire department—which we did, but only because the fire potentially could get to our house. We knew, just as Bill knew, that the sacaton would be fine.

Bill Brophy was the most thoughtful, interesting, and complicated person we met in our years on the Sonoita Plain. He ranched with care and enthusiasm, but without sanctimony. He was never convinced that livestock made all that much difference to the land, but he also didn't think the world was going to collapse if one place went without cows. He was a good neighbor and a friend, and he knew the land and its living things better than most people. An annual highlight for us was touring the Babocomari with Bill once the grasses had greened each summer. Unlike some ecologists, he knew all their names.

Our common ground with Bill Brophy included an appreciation for uncluttered landscapes and unpretentious people. He raised very ordinary livestock, mostly mixed-breed steers imported from Mexico. He was defensive about his motley herd, remarking that "all cattle look the same with their skins off" and "you can't tell one hamburger from another one anyway."

Bill liked a good time, and he had some, but it seemed to us there was a melancholy to him as well. His ranch comprised thousands of acres that were the remnants of a former Spanish land grant—all private property. He once observed that the future of the Sonoita Valley was in real estate and not in ranching. This could have been good news for somebody with that much land, but he wasn't smiling when he said it.

Bill Brophy had it right about lots of things, including fire. He died in 1995, and we miss him.

Seven

OAKS, ACORNS, AND RUGGED GROUPS

THERE IS AN ELEVATED spot next to Turkey Creek on the Research Ranch that affords a fine sweeping view down the drainage and north toward the Mustang Mountains. An old stock tank and a windmill called McDaniel Well occupy the foreground (Fig. 17). Off to the right is a grove of good-sized Emory oak and Arizona white oak trees, whose spreading crowns provide shade and perches for birds and mammals that come to drink.

While most of our workdays were spent in open grasslands on the northern part of the sanctuary, there seemed to be something magical about this wooded place up on Turkey Creek. We came there many times over the years, mostly just to enjoy the view and to share it with others. Although we did not realize it at the time, our initial attraction to that high spot by McDaniel Well was probably far from accidental.

The southern third of the Research Ranch is not a grassland but an oak savanna. The word *savanna* has a complicated history, but it has been applied mainly to the world's tropical grasslands. Especially in Africa, a savanna is a grassland with scattered trees whose wide crowns do not touch, so that a leopard or a monkey has to get down on the ground to move from one tree to the next. Fires frequently burn

FIGURE 17. McDaniel Well in 1998, looking north down Turkey Creek Canyon toward the Mustang Mountains. (Photo by Bill Branan)

through the grassy understory. This, along with the dry climate, helps to explain the low density of trees. African savannas also are the places where our primate ancestors lived and evolved.

The biologist E. O. Wilson coined the term *biophilia* to describe a deep human desire to be in contact with nature. Wilson's biophilia hypothesis asserts that because we evolved in the natural world, we have an emotional connection to it that goes beyond any simple need for its resources.

A fascinating part of biophilia is that there are certain sorts of natural landscapes that most people like more than others. If we are given a choice of places to live, or if we are asked to rank pictures of landscapes, we usually choose a prominent place with a good view, near water, with scattered, broad-crowned trees. These also are the landscapes we build. We like to clear away dense forests, leaving a few of the largest trees scattered around. Among other things, this gives us a better view of our neighbors. If we move onto the prairie or into a desert, the first thing most of us do is plant some trees around our houses.

On the Research Ranch, we were drawn to savannas because of their beauty, and perhaps because they tweaked an ancient memory. But as to the oaks themselves, they attracted us for reasons besides mere aesthetics. First, these were live oaks—evergreen trees that are supposed to keep their leaves all year. Yet in the hot and dry days of May and June, lots of oaks on the Ranch drop their leaves. It is always a patchy sort of thing, with the trees on one hillside nearly all bare and those in the next valley scarcely affected. Visitors in this season often ask us why so many of our oaks have the blight, but that is not it. The trees are just responding to local patterns of drought, coping with the Dead Season, waiting to leaf out with the monsoon.

The other dramatic thing about oaks is their episodic bumper crops of acorns. In the right year a big Emory oak or Arizona white oak can produce thousands of these fat bundles of energy. Animals with a special appetite for acorns include rock squirrels, Mexican jays, javelinas, deer, and acorn woodpeckers. In their collective hunger, these acorn consumers set in motion powerful ecological and evolutionary contests between themselves and the oaks. To a rock squirrel, an acorn is lunch. To an oak, that same acorn is the future.

Relationships between oaks and consumers of acorns are not strictly adversarial. To be sure, a javelina that chews one up has killed an oak seed. But other acorn predators prefer to hoard their food. Some, like Mexican jays, do this by burying the nuts in the ground. If they eventually forget where some of them are, or if they bury more than they need, these acorn predators have inadvertently become oak farmers.

The ambiguous evolutionary relationships between oaks and their acorn predator/planters help to explain three facts. First, individual trees often produce huge numbers of acorns in one year and scarcely any the next. Second, most of the oaks in a local population are coordinated in this, which results in bumper-crop years interspersed with crop failures, a phenomenon called masting. Third, acorns have an unusual chemistry that in fact functions to deter predators.

From an evolutionary perspective, oaks during their lifetimes must get as many acorns as possible past their predators and buried in the ground, where they can germinate and grow up to be the next generation. This is their only legacy. Apparently the best way to do this is to produce a great many acorns, all at once in the fall, so that predators cannot eat them all. It also helps, of course, if some of the acorn predators are acorn planters.

Ecologists have termed this general sort of relationship the predator-satiation hypothesis. An important corollary of the hypothesis is that it is easier to satiate a small population of predators than a large one. This probably explains why oaks do not produce large numbers of acorns every year. Since specialized acorn predators are more likely to starve during crop failure years, there will be fewer around during the next bumper crop year. Ecologists call this the feast-or-famine strategy, and it appears to be a game that oaks play particularly well.

Oaks are conformists when it comes to acorn production, in that powerful natural selection works against the individual tree that makes the mistake of producing acorns when everybody else is taking the year off. Few if any of its acorns are likely to get past all those hungry predators, and even the acorn planters are likely to dig up every single one before the winter is over. All of this leaves open the question of how the oaks might "decide" which year to produce a bumper crop, and which not. It seems that the oaks have evolved common sensitivities to particular sets of environmental signals that trigger flowering and acorn production in certain years and not in others.

In addition to alternately starving and then overwhelming their predators through masting, oaks defend themselves against consumers by lacing their acorns with chemicals called tannins. Tannins are suspected of having three sorts of negative impacts on animals. First, they have a very bitter taste. Second, they disable certain proteins, including digestive enzymes. Third, tannins may directly damage the cells lining the digestive tract. In other words, an acorn meal can give you a spectacular bellyache and more or less lock up your intestines.

To this point we have considered how oaks respond to the presence of acorn predators and acorn planters. But what about the reciprocal side of this ecological relationship? Given the value of acorns as food, it is no surprise that animals have evolved special adaptations to deal with the perverse ways oaks have of serving up their seeds. Simplest among these adaptations is a tolerance for tannins. Most acorns are far too bitter for humans to consume without first leaching out the tannins, but they are readily eaten by specialized acorn predators.

More complex are the ways that acorn predators have adapted anatomically and socially to their unique resources. A good example is the rock squirrel, whose behavioral ecology was studied at the Research Ranch in the 1980s by Joe Ortega. Rock squirrels belong to the genus *Spermophilus,* commonly referred to as ground squirrels because most species in this group are strictly terrestrial. Ground squirrels usually have rather compact bodies and short tails suitable for life spent on and under the ground. Many live in dense colonies in open habitats such as grasslands and deserts, where they feed on foliage, seeds, and insects and seek shelter in underground burrow systems. Ground squirrels often lead complicated social lives, living close to relatives, warning each other about approaching predators, and fiercely defending territories against other groups.

In another evolutionary line are the tree squirrels (genus *Sciurus*). Unlike ground squirrels, tree squirrels usually are solitary and do not defend territories. They also have long, bushy tails and lanky bodies suited to an acrobatic existence in a wooded forest canopy. They frequently have diets specializing in tree seeds, including acorns.

Rock squirrels represent an example of evolutionary convergence. While various aspects of their anatomy and behavior make it clear they are a *Spermophilus,* in many ways they are remarkably like tree squirrels. They have large bodies and bushy tails, and they spend much time in trees. Rock squirrels are as likely to sleep in hollow trunks as in underground dens. Except when the young are dependent on their mothers, or during mating, rock squirrels live alone. Males and females occupy individual but overlapping home ranges, which they do not defend except perhaps in the immediate vicinities of their dens. In spring, they climb to the highest branches of the oaks to eat flowers and leaf buds. In fall, they climb back to pick the surviving acorns.

At least at the Research Ranch, the main reward for a rock squirrel giving up its terrestrial existence is in having access to the oaks and their resources, especially acorns. Higher up in the surrounding mountains there is a *Sciurus* called the Arizona gray squirrel. We have seen one or two on the sanctuary, at the very upper reaches of Lyle and

O'Donnell Canyons. They eat acorns, and they sometimes are willing to travel on the ground. But where the oaks thin out to a savanna they disappear and are replaced by rock squirrels.

Other acorn predators on the Research Ranch tend to favor group living. Most remarkable among these are the Mexican jay and the acorn woodpecker.

Mexican jays probably are the most important acorn planters in the oak savannas of southeastern Arizona, and they must be responsible for a great deal of oak recruitment in this part of the world. In July and August the jays pick acorns by the hundreds and bury them one at a time in the ground, scattered throughout their territories. Each individual acorn is stuck into the leaf litter or soil and then covered with a leaf or a piece of bark or other debris. Mexican jays apparently have remarkable memories, and so most of the acorns get dug up and eaten during the following winter. But many are forgotten or not needed, especially in a bumper-crop year.

Many young trees have been added to Emory oak populations on the Research Ranch since it was founded in 1968. By contrast, there has been relatively little reproduction in the other common species, the Arizona white oak. We do not know if this reflects a preference on the part of Mexican jays for Emory oak acorns, or if it is the result of some other factor. However, it may not be a coincidence that Emory oaks happen to produce acorns low in tannins compared to most of their relatives. Some Native Americans grind them into flour or chop them up and add them to various recipes. We have found bags of Emory oak acorns for sale in markets down in Nogales, Sonora. They are not as good as cashews or almonds, but they certainly are edible. Particular trees happen to produce especially tasty acorns, and these trees, marked with distinctive blazes, can be found scattered through oak savannas on both sides of the Arizona-Mexico border.

No studies at the Ranch have focused specifically on the social behavior of Mexican jays. However, Jerram Brown and his students and colleagues have been engaged in a long-term study of this species east of the Ranch, at the Southwestern Research Station in the foothills of the Chiricahua Mountains near Portal, Arizona.

Mexican jays live year round in groups of five to twenty-five individuals, occupying permanently defended territories. At the start of the breeding season a varying number of males in the group begin to construct separate nests, apparently in an attempt to attract females. Ultimately, two to four females may lay eggs, each in her own nest. All the eggs laid by a particular female might be the result of mating with the male who attracted her in the first place, or there might be several fathers. Males in turn may be the fathers of eggs in more than one nest. In short, Mexican jays lead messy social lives.

The mother alone incubates her eggs and broods her young, but feeding the young both before and especially after fledging is a community activity. Up to half the food delivered to a young Mexican jay likely is brought by somebody other than its parents. These volunteer feeders of both genders are called helpers, and in Mexican jays they may include other breeding adults as well as group members that are not breeding.

Why have Mexican jays evolved such a complicated social system? Why do most of these birds spend their entire lives as members of their natal groups, on their natal territories, even though that means many never get a chance to breed? Why would they bother to feed young that are not their own? The answers to these questions are not entirely clear, but the following factors certainly are involved. First, perhaps largely because of the acorns, Mexican jays achieve high population densities and more or less permanently occupy all suitable habitat. This sort of habitat saturation makes it unlikely that a young jay leaving its natal territory is going to find someplace to live, let alone breed. It does happen, of course, but not very often. Second, if you stay home you end up living with a bunch of relatives—not only your parents, but your aunts and uncles and your older sisters, brothers, and cousins born in previous years. The result is that you are related to nearly all the nestlings hatched in a particular year, and so helping to raise them makes a certain amount of evolutionary sense.

Helping behavior can evolve through a process called kin selection, whereby you increase the frequency of your own genes in future generations by increasing the survival of related individuals carrying many of those same genes. One of the best ways to pass on your genes is to breed yourself, but if that is not possible the next best thing may be to stay home and raise relatives. Finally, whether helping or not, by staying home you may one day inherit a permanent place in the family territory and ascend to breeding status. In the lives of Mexican jays, nothing is more valuable than a place to live.

Another group-living communal breeder at the Research Ranch is the acorn woodpecker (Fig. 18). This species has been well studied in oak woodlands at the Hastings Natural History Reservation in central California, where its social life and relationship to oaks and acorns have been thoroughly documented. Like Mexican jays, acorn woodpeckers live in resident groups of up to twelve relatives, where helpers assist in raising the young. Unlike the jays, there is only one nest per group in a given year. Another important difference, especially for oak trees, is that the woodpeckers store their acorns in specially excavated holes in trees rather than in the ground. These trees and their thousands of acorn-sized holes are called granaries. Granaries are the center of social life for acorn woodpeckers, an invaluable cultural resource that is passed from one generation to the next. If anything, staying home is even more important for acorn woodpeckers than it is for Mexican jays, because without access to a granary an acorn woodpecker is unlikely to survive the winter.

Groups of acorn woodpeckers usually consist of several females that are sisters, a similar cohort of males that are brothers but not related to the female cohort, and varying numbers of young from previous years' nesting attempts. All individuals in the group help to communally harvest and store acorns in the granary in the fall, and all share in this critical resource throughout the winter. Most or all then help to raise the young from the single nest in the spring.

Complex incest-avoidance systems exist such that daughters do not mate with their fathers or uncles, nor sons with their mothers or aunts. The main way this works is that a young acorn woodpecker cannot ascend to breeding status as long as any adult member of the opposite-sex cohort is still alive. When all members of a cohort finally die, another collection of sisters or brothers from a nearby colony is accepted into the group. These are among the most tumultuous events in the lives of acorn woodpeckers, when neighboring groups vie to determine which one gets to provide the new cohort.

FIGURE 18. A solitary acorn woodpecker. (Photo by the authors)

F. A. Leach was one of the first people to study the social system of the acorn woodpecker in California. In a 1925 publication he described it as "communism." Leach had no idea just how complicated things really were inside these groups of birds, but still it wasn't a bad term.

Most of the acorn woodpeckers we studied at the Research Ranch turned out to be capitalists, but not very good ones.

Peter Stacey began to look at these birds on the Ranch in the mid-1970s. There were lots of them around. We could find granaries here and there, but they seemed to be quite a bit smaller than the ones in California. We found that most of the acorn woodpeckers in the oak savannas of the Sonoita Valley lived either as single breeding pairs or alone. Those that attempted to nest usually did so without helpers, and reproductive success was low. The birds harvested and stored a few acorns in the late summer, but they usually left in the fall for parts unknown. Most came back the following spring, but they did not always reestablish on the same territories or with the same mates.

We did find just a few places where Arizona acorn woodpeckers lived in groups of three, and where they remained resident through the winter. These threesomes usually stayed near houses where they had access to backyard feeders and outdoor bowls of dog food.

Our study revealed a level of social plasticity not previously suspected in the acorn woodpecker, and suggested something about the circumstances where communality was likely to happen. One major difference between the oak savannas of southern Arizona and the oak woodlands of California is in the abundance and variety of oaks. Acorn crops of the relatively few oak species in the Sonoita Valley were neither as large nor as reliable as those in California. Without the resources to supply a granary sufficient to get through the winter, acorn woodpeckers adopted a completely different social life. Breeding success seemed to be very low, and we wondered if Research Ranch woodpeckers were operating as population sinks, dependent upon replenishment from their more communal brethren in the nearby mountains or south in Mexico, where the variety of oaks was greater and the acorn crops larger and more predictable. If so, would they shift to the group-living strategy if given the chance?

One of the last things we tried with acorn woodpeckers at the Research Ranch was to experimentally augment their food, to determine if they might become resident and eventually communal. We found an old fallen telephone pole, drilled it full of holes, and loaded it up with hundreds of Emory oak acorns purchased in the markets down in Nogales. We dug a hole in the ground next to a particular pair's nest tree and then, in the middle of the night, set up the instant granary.

In the morning, the acorn woodpeckers awoke to discover that a large and very magical object had mysteriously appeared right next door. It reminded us of the struggling apes encountering the black monolith in Arthur Clarke's *2001: A Space Odyssey.* But the woodpeckers did not develop a new and higher intellect, or travel to the moon, or go into a psychedelic orbit around Jupiter. Instead, they unloaded the acorns as quickly as they could, ate them all up, and then headed off for Mexico as usual. Maybe the socialization process in acorn woodpeckers requires building the granary themselves and not just being handed one.

Collared peccary (also called javelina) are characteristic animals of the Research Ranch, and we studied their movements in relation to season and habitat as well. While javelina are not strictly acorn specialists, in the fall we found many of their scats filled with little else but pieces of acorn shells.

Mostly nocturnal during summer, these New World equivalents of wild pigs sought shelter during cold winter nights in heavy cover such as sacaton grass and then foraged for plant tubers, agave, cacti, acorns, and other plant foods during the sunny winter days.

Javelinas are very tough animals, able to chew up cacti, spines and all, with no apparent ill effects. They find underground foods using a keen sense of smell, anchor themselves with their short, stubby legs, and then dig out tasty morsels with their leathery snouts, like miniature bulldozers. They tear apart agave plants with their tusks and hooves and then eat the leaf bases like artichokes. We have seen animals walking casually across the Research Ranch, apparently unaware of cactus pads and swordlike agave leaves stuck in their hides.

For all their durability, javelinas thrive only in tightly integrated groups of up to a dozen or more individuals. Almost any sort of behavior is tolerated within the group, especially from the communally raised young. Juveniles can follow any foraging adult around and snatch food from its mouth without retribution, and they are welcome to nurse any lactating female. Adults fiercely and cooperatively defend one another, and especially their young, against predators such as bobcats and coyotes. Solitary javelinas do not live long.

Nearly all social interactions within a javelina herd are benign, but between groups they are aggressive and territorial. In the early days, ranchers sometimes caught young javelinas and kept them as pets. They would grow up to identify with their adopted human families and defend them against the mailman, visiting distant relatives, or any other unwelcome stranger. The success of the javelina in the American Southwest is testimony to the advantages of group living in extended families.

Former U.S. senator Barry Goldwater died on May 29, 1998. One obituary commented that he exemplified the spirit of rugged individualism characteristic of life in the American West. It diminishes neither the senator's influence, nor his unqualified love for Arizona, to point out the folly of such a notion. Hardly anybody important, least of all an effective politician, operates alone. If Barry Goldwater had spent all his time in a lonely line shack tending fence out on the north forty, he would probably have been a rugged individualist, but none of us would recognize his name.

Our ancestors survived, evolved, and eventually prospered in those ancient African savannas not because they were particularly rugged, and certainly not because they were bigger or stronger or sneakier or faster than the competition. They were none of those things. They made it because they learned how to communicate and cooperate. Through cooperation they discovered ways to find food without becoming food themselves. Through language they passed on their accumulated discoveries to subsequent generations, at blinding speeds compared to the trudging (if inexorable) pace of evolution.

The strategy was so successful that our ancestors took it with them and colonized the rest of the globe. We have now become the dominant life force on our planet, operating not as rugged loners but as groups. As a profoundly social animal now living mostly in crowded places, it is only in cooperative and inclusive groups that we have any chance of finding ways to sustain ourselves and the rest of the natural world over the long term. The alternative is anarchy.

Eight

LITTLE BROWN BIRDS

Sparrows are among the dullest and most enigmatic of birds. Nearly all are cryptically patterned in subtle shades of brown and gray, and they spend most of their lives on the ground, quietly searching for seeds and insects while avoiding being eaten themselves. They are conspicuous and distinctive only to the ear and then only at breeding times, when they sing to proclaim their territories and attract mates. Yet sparrows can be powerful indicators of environmental condition, and they have much to tell us about the ecology of the grassland communities of which they are a part. More than twenty species occur with some regularity on the Research Ranch, each having something unique to say about life and times in the Sonoita Valley. Telling all their stories would exceed the limits of this book as well as the tolerance of most readers, so what follows might come under the general heading of "Sparrow Highlights."

AMMODRAMUS

Grasshopper sparrows (*Ammodramus savannarum*) are common year-round residents of upland grasslands on the Sonoita Plain (Fig. 19). These small, compact birds have short wings and tails compared to

FIGURE 19. Grasshopper sparrow. (Photo by the authors)

most others of their kind, and they have distinctive flattened foreheads. They fly from one grass clump to another with characteristic rapid and shallow wing beats. Grasshopper sparrows are so called not because of what they eat, but because their songs resemble the raspy buzz of an insect. However, it also turns out that they are major predators on grasshoppers, and they may have a controlling influence on abundance of some of their prey (more on which later).

In the early 1980s we counted numbers of grasshopper sparrows in summer and in winter, both on and adjacent to the Research Ranch. These birds were more than seven times more abundant on the sanctuary than off, indicating a clear preference for one sort of habitat over the other. What was the basis of this preference? We can never know for certain, because there is no way to get inside the heads of these little birds. Short of that, our approach to understanding the world as seen through the eyes of a grasshopper sparrow was to measure attributes of the places they chose to live.

In the summer of 1982 we marked central places on the breeding territories of thirty-five pairs of grasshopper sparrows. We then measured characteristics of vegetation around these points, as presumptive indicators of the birds' habitat preferences.

The average song perch of a nesting grasshopper sparrow was centered on a patch of grassland with the following attributes: 72 percent grass canopy, 4 percent herb canopy, 5 percent shrub canopy, and an average of 23 percent bare ground.

These numbers describe a typical grasshopper sparrow territory, but by themselves they still did not tell us why the birds were so much more common on the sanctuary than on adjacent operating cattle ranches. To ascertain this, we had to measure vegetation characteristics at random points both on and off the sanctuary and then compare these with the data from the grasshopper sparrow territories.

Table 2 reveals several interesting things about the vegetation of grasshopper sparrow territories compared to our sample sites. Most striking is the amount of grass cover, which was an average of 20 percent higher than at random points on the sanctuary and twice as high as on adjacent grazed lands. Small wonder that grasshopper sparrows were so rare off the Research Ranch. Even on the sanctuary, they were selecting patches of grassland with exceptionally dense grass cover.

Another important point to understand about grasshopper sparrow habitat selection is that these birds are not actually avoiding livestock. We know this because in other parts of North America they are common where livestock occur. For example, in the prairies of the eastern Great Plains, grasshopper sparrows are more typical at grazed than ungrazed sites. In these lush and highly productive ecosystems, apparently, bare ground does not occur in sufficient quantity except in the presence of grazing. In other words, the same ecological force that destroys grasshopper sparrow habitat on a southwestern plain actually creates it in the tallgrass. This illustrates why conservation plans must be site- and species-specific, and why sweeping generalizations about what is or is not good for wildlife can get us into trouble.

One of the remaining uncertainties about grasshopper sparrow biology is how they actually operate on the ground. We suspect that they are ambush and stalking predators, hiding in grass clumps until an

TABLE 2. Percent vegetation canopy on the territories of grasshopper sparrows and at nearby random points on the sanctuary and on adjacent grazed land

		Random Points	
	Territories	*Sanctuary*	*Grazed Land*
Percent grass cover	72	52	36
Percent herb cover	4	13	10
Percent shrub cover	5	6	1
Percent bare ground	23	27	51

insect passes by or moving slowly through the heavy cover until they come upon an unsuspecting prey. They may also be dependent on grass canopy to hide from their own predators. Whatever the reason, their quiet and furtive behavior makes them virtually undetectable when they are engaged in the essential business of finding food for themselves and for their young.

There is another bird in the genus *Ammodramus* that visits the Research Ranch, and if anything it is even more secretive than the grasshopper sparrow. The Baird's sparrow (*A. bairdii*) is a rare and narrowly distributed species that nests primarily in grasslands of North Dakota and the Canadian Prairie Provinces. In fall it migrates south to grasslands of extreme southeastern Arizona and northern Mexico. Baird's sparrows winter on the Research Ranch, along with the grasshopper sparrows, and it is very difficult to tell them apart.

Ron Pulliam is an internationally renowned ecologist who worked extensively on the Research Ranch through the 1970s, developing and testing theories about the ecological and evolutionary forces that determine the organization of bird communities. He once shared with us his "secret" for distinguishing grasshopper from Baird's sparrows in the field in winter. If you repeatedly flush a small short-tailed and short-winged bird out of the grass and it eventually sits up to get a look back at you, it is a grasshopper sparrow. If it never comes up to look at you, says Ron, it is a Baird's sparrow. Subsequently, Caleb Gordon captured *Ammodramus* sparrows on the Research Ranch using fine-mesh nets. It turns out that Baird's sparrows are much more common than anyone previously suspected, which strongly suggests that this rare bird may be very much dependent on ungrazed grasslands in winter.

Ron Pulliam moved on to bigger things before the National Audubon Society assumed responsibility for the Research Ranch in 1980. Among other positions, he was the first director of the newly created National Biological Service (subsequently the Biological Resources Division of the U.S. Geological Survey). This federal agency has the challenging job of monitoring and publicizing information about the distribution, abundance, and status of our nation's wild flora and fauna. Conservation of our natural heritage relies critically on such information, but this vital agency has been under frequent attack since its inception, the victim of a campaign of distortions brought by powerful forces that recognize all too well the importance of killing the messenger in any environmental debate.

So, a career in sparrows is not the professional blind alley you might suspect, but it can get you into the eye of a hurricane. We wonder if Ron Pulliam misses the Sonoita Plain.

AIMOPHILA

Some other grassland birds in the Sonoita Valley actually make the grasshopper sparrow look garish by comparison. Two of these are the Botteri's sparrow (*Aimophila botterii*) and its very close relative, the Cassin's sparrow (*A. cassinii*). Both are relatively long tailed and long winged, but otherwise they are virtually lacking in good visual field characteristics (Fig. 20). The Botteri's sparrow is a nearly uniform grayish brown and has been described as being "as plain as a mud fence." The Cassin's sparrow is only a slightly grayer version of the same thing.

Both Botteri's and Cassin's sparrows prefer very substantial grasslands, and both declined in Arizona following the increases in livestock late in the last century. Botteri's sparrows went unrecorded in the state from 1903 to 1932, and as recently as the early 1960s no nests of either species had been discovered in the state. Today, however, the Botteri's

FIGURE 20. Botteri's (left) and Cassin's (right) sparrows. (Photo by the authors)

and Cassin's sparrows are two of the most abundant nesting birds on the Research Ranch.

Both these species have conspicuous and characteristic songs. This is fortunate, because their habits and habitats turn out to be very interesting and very different, and it might not have been possible to sort them out had we not been able to distinguish the singing males. The territorial song of the male Botteri's sparrow is a short, slow-paced trill, often preceded and followed by two or more separate notes. The rhythm of the trill has been likened to that of a bouncing ping-pong ball coming to rest, accelerating in tempo as it goes along. Botteri's sparrows usually sing from a perch, but they sometimes sing in flight as well.

The Cassin's sparrow's song is much more complex, consisting of a rapid, more prolonged, and somewhat plaintive trill, preceded and followed by different numbers and types of notes, the last one of which almost always is distinctly higher in pitch than any that have come before. Cassin's sparrows frequently sing in flight, often as part of stereotyped arching and fluttering "song-flights" that carry the birds from one territorial perch to another.

Cassin's sparrows nest from the plains grasslands of eastern Colorado and southwestern Nebraska, south into Texas, southern New Mexico and southeastern Arizona, and on into grasslands of northern Mexico. They winter in the southern part of the breeding range. Within this large area, they are perhaps the most enigmatic of all grassland birds.

The Cassin's sparrow is a wandering opportunist seeking resource hot spots that come and go unpredictably across both time and space. Like all opportunists, the Cassin's sparrow is tough and adaptable, but it is absolutely dependent on finding places where winter seeds or summer insects are abundant. The only other consistent feature of their habitat is an apparent requirement for comparatively lush grasslands with scattered shrubs or small trees from which the males can launch their song flights. It is rare to find good numbers of these birds, except in places that are ungrazed or lightly grazed. However, they are remarkably good at finding even the smallest and most isolated of these habitat islands.

The high desert country along the Rio Puerco in northern New Mexico is a stark landscape of mesa, rim rock, and broad sandy plains. It also is one of the most thoroughly desertified pieces of the American public land trust, due to its inherently unstable soils and to centuries of livestock grazing. We were counting birds up in this country one summer when we came upon a small livestock exclosure, less than ten acres in extent, in the midst of a broad valley of sand and desert scrub. The country as a whole was best suited to jackrabbits and horned larks, but nesting in this tiny island of lush grassland was a small population of Cassin's sparrows, isolated by dozens if not hundreds of miles from their nearest neighbors.

Through the 1980s Cassin's sparrows were more than ten times more common on the Research Ranch than on adjacent grazed lands of the Sonoita Valley, but even here their numbers were variable and unpredictable. Usually summer birds arrived on the sanctuary in June and began nesting in July with the onset of the summer monsoon. In most, but not all, years a second wave of apparently migratory birds would come in from somewhere during August. It was not absolutely clear that these late-summer migrants actually nested, but the males certainly established territories and performed song flights in front of females. Cassin's sparrows also wintered on the sanctuary, but not every year. Furthermore, there was no particular correlation between winter abundance and the numbers of birds that had been present the previous breeding season.

For all their opportunism, Cassin's sparrows appear to be in trouble. Breeding bird survey data indicate this adaptable species is declining throughout much of its range—likely testimony to the ubiquity and homogenizing impacts of livestock grazing on their preferred habitat.

The first detailed study of the breeding biology of the Botteri's sparrow was conducted between 1981 and 1983 on the Research Ranch, by Betsy Webb, and what follows is based largely upon her work.

Botteri's sparrows spend the winter in Mexico and return to their breeding sites in Arizona during the middle part of May. Most apparently come in search of a very particular sort of nesting habitat, at least on and near the Ranch, and this habitat specificity probably explains both their historical decline and continuing low numbers in the American Southwest.

On the Sonoita Plain, breeding Botteri's sparrows usually are associated with stands of sacaton grass. Sacaton grows in tall, dense stands on broad and gently sloping outwash floodplains associated with major stream and river drainages in the region. As we described in Chapter 6, watershed destruction, flash flooding, livestock overgrazing, and resulting erosion have eliminated most of these floodplains, reducing sacaton to a small percentage of its historical range. Some of the very best remaining sacaton stands occur on the Sonoita Plain, and particularly on the Research Ranch, where Botteri's sparrows are unusually abundant. It is not an accident that Betsy Webb chose the Ranch as the site for her work.

Even within sacaton habitat, Botteri's sparrows are picky about where they breed. On and near the Ranch we found that the birds were common only in mature, unburned, and ungrazed or lightly grazed stands, and only where these stands were adjacent to intact upland grasslands. The birds usually built their nests on the ground under the overhanging leaves of a sacaton clump growing near the edge of the floodplain. Breeding territories included a good-sized piece of adjacent upland grassland, where the birds foraged for food (mostly grasshoppers) for themselves and their young. Off the Ranch we have found good sacaton stands completely devoid of Botteri's sparrows, apparently because the adjacent uplands had been desertified from grassland to scrub vegetation.

Botteri's sparrows arrived at the Ranch in May, but they did not begin to breed until the onset of summer rains, usually in early July. In fact, females generally began to lay eggs immediately following the first heavy rain, suggesting that rainfall was the trigger for reproductive activity. Females then laid one egg per day, until the clutch of two to four eggs was complete. Incubation began with the laying of the final egg, and ten days later the eggs began to hatch. Both parents fed the nestlings for another twelve days, at which point the young began to leave the nest. Fledged juveniles remained with their parents for several weeks, but by mid-September nearly all birds had departed for their wintering grounds in Mexico.

The two most interesting aspects of the Botteri's sparrow breeding season are its timing and its brevity. The birds seem to be wedged between two conflicting environmental requirements: finding sufficient food for their young and avoiding floods. The flood danger is low at the beginning of the summer, but then there is less food. As insects increase in response to the greening of vegetation, so does the likelihood of a flash flood. Natural selection appears to have favored Botteri's sparrows that wait for the rain but then complete their nesting cycles in the remarkably short period of twenty-four or so days, from laying of the first egg to fledging of the last young. Even so, in some years the birds have a tough time.

Botteri's sparrows seem to have put themselves in harm's way by choosing sacaton floodplains as nesting habitat. It is worth pondering why they have done so, from both an ecological and an evolutionary perspective. It cannot be for food, because they frequently forage outside the sacaton. The two most likely benefits of nesting under a large grass clump are the favorable thermal environment and concealment of the nest from predators. But why do it in sacaton? A longer historical perspective, and some recent data from the Research Ranch, may answer the question.

Based on historic accounts, we strongly suspect that the summer flash floods which characterize most of the drainages in southeastern Arizona are a consequence of watershed destruction by grazing. Prior to the introduction of livestock, the infiltration and storage capacities of most watersheds were such that the streams flowed relatively gently year round, rather than seasonally and violently as they do today. In other words, the flood risk to Botteri's sparrow nests, especially near the edges of sacaton stands, must have been much lower in the past.

One of the most tantalizing aspects of the life history of the Botteri's sparrow on the Research Ranch is that this small bird appears to be expanding its breeding habitat beyond sacaton and into the uplands. Through the 1970s and into the early 1980s it was hard to find these birds anywhere outside sacaton floodplains. In the mid-1980s, however, we began finding these birds nesting in ungrazed upland grasslands. Though only about one-third as common as grasshopper sparrows, Cassin's sparrows, or meadowlarks in these sites, they were the next most abundant breeding bird.

Could it be that the Botteri's sparrow once was a bird of the uplands at least as much as of sacaton floodplains? Could it be that sacaton stands which survived events of the 1890s were the only suitable habitat remaining? Perhaps only here was the grass cover sufficient to provide the requisite thermal conditions and predator protection. Could the nearly obligate association between Botteri's sparrows and sacaton itself be an historic artifact? Evidence from the Research Ranch suggests that this may be so.

The lesson of the Botteri's sparrow is as important as it is commonplace. This bird did not put itself in harm's way. We did.

SPARROWS IN WINTER

Each fall, millions of sparrows depart their breeding grounds across North America and head south for the winter. They leave to escape the snow cover that denies them access to their principal foraging substrate—the ground. They come in search of their principal winter food—seeds. Circumstantial evidence indicates that most sparrows travel only as far as is necessary to accomplish both goals: that is, to find a good seed source not covered by snow.

In the West, most sparrows apparently travel south until they reach a grassland that suits their habitat requirements and that happens to have produced a sufficient quantity of seeds. In some years this can be in Arizona, or in New Mexico or Texas. In other years the birds must travel on, moving down into the grasslands of northern Mexico. Barny Dunning and Jim Brown studied the regional abundance of sparrows wintering in southeastern Arizona and found it was powerfully correlated with rainfall from the previous summer. At the Research Ranch, Ron Pulliam and his coworkers demonstrated that rainfall strongly influenced grass seed production. In wet years, as many as 21 million grass seeds per acre could await the winter sparrows. After a dry summer, in contrast, as few as 57,000 grass seeds per acre might be available; in these years comparatively few sparrows wintered on the Sonoita Plain.

A dozen species of ground-dwelling and seed-eating songbirds winter regularly on the Research Ranch. Six of them are migratory and come only in winter, another five are resident year round, and one, the Cassin's sparrow, is opportunistic and somewhat unpredictable in both seasons. These twelve species divide up the habitat in generally predictable ways regardless of their numbers from one year to the next. The major habitat gradient that separates them is the type and amount of cover into which the birds can flee when they are disturbed. Table 3 lists the twelve species, ranked in order from those associated with comparatively dense cover to those typical of the most open grasslands.

Whenever ecologists see a pattern like this, they naturally seek to explain it. Why does the chestnut-collared longspur forage only in the most barren sites, far from any shrub or tree cover? Why does the white-crowned sparrow restrict itself to feeding directly under the canopy of tall and dense streamside shrubs?

Ron Pulliam, Scott Mills, Steve Lima, Tom Valone, and various other first-rate avian ecologists have examined the spatial arrangements of sparrows wintering on the sanctuary, and they have arrived at two

TABLE 3. Habitat associations of twelve grassland birds wintering on the Research Ranch

Species	*Major habitat*
white-crowned sparrow	heavy brush cover along drainages
rufous-crowned sparrow	shrubby ravines and tall grasslands
canyon towhee	riparian woodlands and shrublands
chipping sparrow	oak woodland
vesper sparrow	shrubby grasslands
Savannah sparrow	shrubby grasslands
Cassin's sparrow	shrubby grasslands
eastern meadowlark	grasslands with variable cover
grasshopper sparrow	open grasslands, heavy grass cover
Baird's sparrow	open grasslands, heavy grass cover
horned lark	open grasslands with bare ground
chestnut-collared longspur	open grasslands with bare ground

different but complementary explanations for the patterns they found. First, the spatial segregation of sparrows may be the result of competition for the limited seed supply they all share. Perhaps each species, as a result of its unique morphology, forages most efficiently and successfully in a particular habitat and aggressively keeps other species away, hence the abundance of that species in that habitat. A second possibility is that the birds are choosing those habitats where they are best able to avoid being eaten by predators. Prairie falcons, Cooper's hawks, American kestrels, loggerhead shrikes, and other avian predators are common in the area, and all feed substantially on birds. Surviving the winter in the Sonoita Valley is a matter of finding food while not becoming food.

Careful observations of how individuals of each species react to intruders, combined with some ingenious field experiments, indicate that the different species are choosing to forage in those places where each can most efficiently avoid becoming someone else's dinner. For example, longspurs and horned larks flee predators by flying off rapidly, a tactic that works best in very open country where the predators are likely to be spotted at a distance. Grasshopper and Cassin's sparrows, in contrast, escape by diving quickly into heavy grass cover. It is not surprising that these two birds are so much more common on the sanctuary than on adjacent grazed lands, where heavy grass cover is relatively scarce.

Most of the other grassland species flush to shrub or tree cover when disturbed, and they seem very reluctant to stray very far away from it. Steve Lima and Tom Valone tested the importance of this cover experimentally on the sanctuary. They cut branches off mesquite trees and piled them up in areas of the Ranch that formerly had been relatively shrub-free grasslands. Before the manipulations, horned larks and grasshopper sparrows were the common species. After the mesquite shrub piles were in place, grasshopper sparrows and horned larks left, while vesper sparrows became the most common birds. Even a few chipping sparrows showed up—a surprising result, since this species rarely strays far from oak woodlands on the sanctuary.

Experimental additions of cover created new habitat for sparrows unwilling or unable to forage more than a few yards from a shrub or tree into which they could escape when disturbed. But why did these additions also result in the departure of birds not dependent upon such cover? An obvious explanation is that they were forced out by aggressive competitive interactions. However, the evidence for such a relationship among Research Ranch sparrows is equivocal, and more experimentation needs to be done. Are Research Ranch sparrows mostly watching the ground and one another, or are they mostly looking over their shoulders? The answer seems to be: probably both.

SPARROWS AND GRASSHOPPERS

In the summers of 1983 and 1984 Karen Jepson-Innes compared grasshopper populations on and adjacent to the sanctuary to determine how this dominant and ecologically important group of grassland insects differed between grazed and ungrazed areas. She found that most grass-feeding species of grasshopper were more common on the sanctuary, whereas the grazed pastures were dominated by grasshoppers that fed more on broad-leaved herbs and that showed a stronger preference for breeding and basking on bare soil.

Incidental to her other work, Karen built an observation blind near a Cassin's sparrow nest on the Ranch, so close that she could sit inside it and identify the insects the parents were feeding to their young. In eighteen hours of observation she watched the male and female bring in 208 insects. Of these, 197 were grasshoppers. We also had available some general data about the densities of both grasshoppers and nesting songbirds in the area. Taken together, these preliminary results suggested that birds at least had the potential to be affecting abundances of their grasshopper prey. In the summer of 1987 we initiated a four-year experiment to test this hypothesis. Results illustrate both the challenges and the promise of experimental field ecology.

One of the oldest ideas about farm and city birds is that they perform a valuable service by preying on the insects that otherwise would ravage our crops, yards, and gardens. While birds are under no obligation to so justify their existence, nevertheless this idea has some profound ecological implications. Can songbird populations actually exert a controlling influence on their insect prey? Would it matter to the structure or function of their ecosystems if all the birds simply disappeared? This hypothesis has not been tested very often, probably because of the ethical and logistic challenges of performing the necessary field experiments.

Ecologists at one time assumed that songbirds had relatively little ecological impact, especially in grasslands, because their numbers were so low and their prey so abundant. There seemed to be countless thousands of insects out there, and not very many birds to do the eating. Now we have some good experimental evidence that birds can be effective and important predators in some grasslands, on some insect groups. Part of this evidence, thanks to the initial curiosity of Karen Jepson-Innes, comes from the Research Ranch.

In June 1987 we established a permanent grid of forty-eight rectangular study plots on the North Mesa, each 30 by 15 meters (Fig. 21). We estimated densities of grasshoppers on each of these plots. The counting technique involved placing wire hoops on the ground inside the boundaries of each plot. We then sneaked up on each hoop and

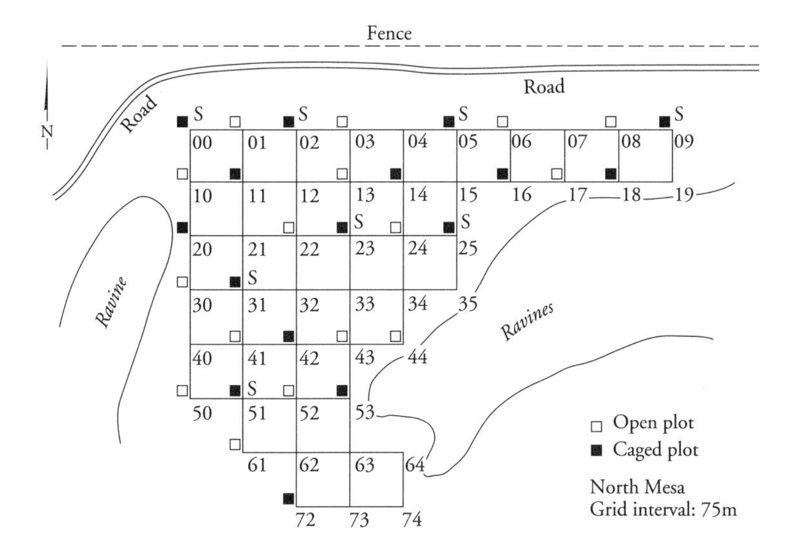

FIGURE 21. Map of grasshopper sampling plots on the North Mesa of the Research Ranch. Caged plots labeled "S" also were fitted with ground-level window screen barriers. (Redrawn from C. E. Bock, Bock, and Grant 1992)

counted and identified the grasshoppers as they jumped out. This is a standard grasshopper census method that works well in areas of relatively low and sparse vegetation and where the variety and abundance of grasshoppers is not too great.

The purpose of the 1987 sampling was twofold. First, we needed baseline data on grasshopper abundance in plots to which birds had full access, before we began the experimental manipulations. Second, it is widely known that grasshopper densities can be highly variable across very small spatial scales, so chance variations among the plots might have made it impossible to detect statistically significant changes resulting from bird exclusion. Therefore, at the end of the 1987 field season we eliminated from further study those sixteen of the forty-eight plots with the highest and lowest overall grasshopper densities. This left sixteen control and sixteen experimental plots, with comparable grasshopper densities.

Over the winter of 1987–1988 we constructed cages over the sixteen experimental plots, designed to exclude birds but allow the free move-

FIGURE 22. Jane Bock in 1989 beside bird exclosure on the North Mesa used to determine the impacts of bird predation on grasshoppers. (Photo by C. Bock)

ment of grasshoppers in and out (Fig. 22). Each cage was two meters high and covered the entire 30×15–meter plot. The walls were chicken wire, and the roofs were pieces of mesh netting manufactured mainly for farmers who need to keep birds off their berry patches.

Through the summers of 1988, 1989, and 1990 we counted grasshoppers inside the sixteen North Mesa bird exclosures and compared their densities with those in the sixteen uncaged plots. Each plot had twenty-eight hoops, and grasshoppers were counted about once every ten days from May through September of each year. Over the full four-year duration of the study, we ended up censusing grasshoppers in individual hoops 44,208 times.

Recall that grasshopper densities did not differ between control and experimental plots prior to cage construction in 1987. Differences between the caged and open plots appeared the first summer the cages were in place (1988), and these differences increased each subsequent year. By 1990, overall abundances of grasshoppers on the plots protected from bird predation were more than double those where birds were present. Seven of the twelve most common grasshopper species were significantly more abundant inside the bird exclosures, while none were more common outside.

Although these results seemed clear and dramatic, there were at least two ways they could have been artifacts of our experimental design. First, the increased grasshopper abundances on the experimental plots could have been due to changes in cage microclimate. Perhaps the mesh ceilings and walls increased shading or reduced temperatures and wind, thus creating environments more favorable to grasshoppers, vegetation, or both. Second, perhaps the grasshoppers somehow recognized the cages as predator-free refuges and colonized them in large numbers from the outside. Given the small size of the caged plots relative to the whole of the North Mesa, even a slight tendency for grasshoppers to move into the cages could have resulted in much higher numbers inside than out, unrelated to direct differences in survival from predation.

We took steps to investigate both of these possible artifacts, and we concluded that neither was a likely explanation for differences in grasshopper densities inside versus outside the caged plots.

First, we could find no dramatic microclimatic differences inside the cages for such variables as light intensity, wind speed, temperature, or precipitation. Furthermore, plant species composition and overall vegetation cover did not differ between caged and open plots.

Second, in both 1989 and 1990 we fitted the perimeter of half of the sixteen caged plots with a one-meter-high barrier of fine-mesh window screen, anchored to the ground and attached to the chicken wire walls. While these ground-level window-screen fences certainly were not absolute insect barriers, we reasoned that they would substantially retard movements, since most dispersal in grasshoppers occurs by walking rather than by flying or jumping. What we found was that densities of grasshoppers became significantly higher in the cages with the window screen barriers than in those cages without the screen or on the uncaged control plots. This suggested that the experimental plots were sources of grasshopper emigration, rather than recipients of grasshopper immigration. If anything, the insects were attempting to escape crowding inside the bird exclosures.

We took the bird exclosures down in 1991, and we were sorry to see them go. But intrusions such as these should not be permitted to accumulate on the Research Ranch, lest they compromise its higher purposes.

Similar field experiments conducted elsewhere in North America have shown that sparrows and other songbirds can limit grasshopper densities in many, but not all, grasslands. At least on the North Mesa of the Research Ranch, the little brown birds have a profound ecological influence on their food supplies. Not only are they an aesthetic part of the Sonoita Plain, and individually powerful indicators of environmental condition, but they appear to be key players in the dynamics of these grasslands as well.

Nine

FRAGMENTS

A Wright's sycamore stands tall and absolutely alone in the middle of O'Donnell Canyon below East Corrals (Fig. 23). In 1981 it became riparian tree #70, thanks to a small aluminum tag we nailed to its trunk. Of course, this sycamore had survived for centuries without the benefit of a number, perhaps even since the time of Coronado. Somehow it had endured the flash floods that have scoured O'Donnell Creek over the years and the fires that have burned through the adjacent sacaton. But southwestern riparian—or streamside—habitats as a whole have suffered grievously from historical changes in the watersheds they drain, and we were interested to learn how their tree populations have fared on the sanctuary and elsewhere in the Sonoita Valley. One way to do this was to tag as many trees as possible, to follow their fates, and to look for new recruits.

Other riparian trees—Fremont cottonwood, velvet ash, Arizona walnut, desert willow, and true willow—have added numerous youngsters to their populations on the Research Ranch since 1968. But the sycamores are having a hard time of it. Their seedlings establish well only in the sandy soil of bare creek bottoms, where they almost invariably get washed out by floods during the summer monsoon. The good news is that adult sycamores seem to be virtually immortal once

FIGURE 23. Sycamore #70 in 1998. (Photo by Bill Branan)

their root systems grow large and strong enough to tolerate the floods. Sycamore populations are hanging on. They are even recruiting a few new individuals in those limited places in southeastern Arizona with a reliable water source, and where the watersheds happen not to flood very often.

Early on the morning of May 16, 1984, Tom Strong walked down to sycamore #70, stood under it for five minutes, and counted all the birds he could see or hear in its branches and in the sacaton grasslands immediately adjacent. He repeated this ritual seventeen more times during the summers of 1984 through 1986, not only under that particular tree but at 131 additional riparian spots around the region. He counted birds in towering cottonwoods along Babocomari Creek, in gnarled walnuts in Vaughn Canyon, in desert willows in Blacktail Canyon, in shaded groves of big-tooth maples deep in Garden Canyon, in sycamores high in the pine-oak woodlands of Sawmill Canyon, and in a dozen other riparian circumstances in the Sonoita Valley and east up into the mountains on Fort Huachuca.

Riparian woodlands make up less than 1 percent of the landscape of the American Southwest, yet they support an extraordinary abundance and variety of birds and other wildlife. They had been well studied in big river bottoms down in the desert, especially along the Lower Colorado in western Arizona. But much less was known about the importance of riparian woodlands to birds in the basin and range country of southeastern Arizona. How important were these woodlands to birds in this part of the world? Were certain sorts of riparian habitats more important than others, in terms of the bird populations they sustained? With sponsorship of the U.S. Forest Service and cooperation of the army at Fort Huachuca, we began a study designed to answer these questions.

Specifically, we asked four things about riparian bird populations in our part of the world. First, were there differences depending on which riparian tree species dominated a particular site? Second, were there differences depending on whether a particular drainage flowed through grasslands, such as on the Sonoita Plain, or through woodlands, such as up in the Huachuca Mountains? Third, did the size of the riparian woodland matter? Finally, were riparian trees even necessary to attract birds to drainages? To answer the last question—and as a sort of control for our experiment—several of our plots were located along washes in places where there happened not to be any riparian trees at all.

Results of the study were interesting and complex. The big picture included three things, two of which were not in the least surprising and the last of which we did not expect.

First, stands of the largest riparian trees, especially Wright's sycamore and Fremont cottonwood, supported a greater abundance and variety of birds than the smaller trees. Cottonwoods and sycamores not only had much more foliage for insects (bird food), but the oldest and largest individuals usually had hollow limbs and trunks—critical spaces for cavity-nesting birds such as woodpeckers, wrens, titmice, nuthatches, bluebirds, and certain kinds of flycatchers.

Our second finding was that riparian trees of any sort were much more important to birds in grasslands than they were in forested landscapes over in the Huachuca Mountains. In fact, control plots—where there were no riparian trees—embedded in forested ecosystems scarcely differed from plots that had riparian trees growing on them in terms of the overall density or variety of birds. Apparently an oak or a pine or a large juniper next to a montane stream served birds nearly as well as a maple or a cottonwood or a velvet ash in the same circumstances.

The result we did not expect had to do with the sizes of riparian woodlands in grasslands, and this brings us back to sycamore #70. Over the three years, Tom Strong counted thirty-five bird species in or near this solitary tree, at a combined average density of more than thirty birds per hectare (about 2.5 acres). Some of the common ones were typical riparian nesters, such as the American kestrel (a small falcon), northern flicker, Cassin's kingbird, Bullock's oriole, and black-headed grosbeak.

Downstream toward the boundary between the Research Ranch and the Babocomari was control plot #563. It had no riparian trees, although there were a few small mesquites scattered among the grasses. Tom found only twenty-two species on plot #563, with a combined density of fewer than four individuals per hectare. There were some Botteri's sparrows and common yellowthroats, both of which nest in sacaton grasslands. But there were no kestrels or flickers or grosbeaks or orioles, and fewer than one-fiftieth as many Cassin's kingbirds. In other words, sycamore #70 was critically important to lots of the birds nesting in Lower O'Donnell Canyon.

The outwash of Garden Canyon over on Fort Huachuca looked to us a lot like Lower O'Donnell Canyon on the Research Ranch: a broad floodplain grassland and some scattered mesquite, with a meandering (usually dry) creekbed running down the middle. The big difference had to do with Wright's sycamores, which lined Garden Canyon by the dozens (Fig. 24). Plots 587–592 were located here, scattered over a two-mile stretch of the canyon, and each one included several large sycamores. When our project was done and we added up the numbers, it turned out that each of the six Lower Garden Canyon plots supported an average of thirty-six bird species, at an average combined density slightly lower than that of sycamore #70. In other words, at least on a local scale, one big sycamore was just as good for birds as a long string of them.

FIGURE 24. A sycamore forest in Lower Garden Canyon, 1998. (Photo by the authors)

The reason this result surprised us is that it ran counter to an emerging principle of the young science of landscape ecology. Landscape ecology was born at Harvard, and some of its most rigorous and successful applications have involved studies of birds associated with woodland habitats, and to a lesser extent grasslands, in the northeastern and midwestern parts of the United States. At the core of landscape ecology is the hypothesis that the kinds and abundances of species associated with a particular patch of habitat depend not only on the nature of the patch but also on its size and on the sorts of habitats surrounding it—in other words, on its landscape context.

Landscape ecological studies have taught us that a small woodlot isolated in an agricultural landscape of Iowa farmhouses and cornfields, or in a suburban Pennsylvania neighborhood, is likely to have very different birds in it than a larger, more continuous woodland. This will be true even if the smaller patch has all the trees and shrubs and understory grasses you would find in a bigger patch, and even if it is just as able, in terms of sufficient size, to accommodate individual breeding pairs of birds.

The principles of landscape ecology have profound management implications for the conservation of birds and other wildlife. Remaining habitat patches in fragmented landscapes such as midwestern woodlots must be above a certain minimum size or they are likely to be missing whole groups of species that for various reasons need larger tracts of suitable habitat.

We are only beginning to understand why some species may be rare or absent from small remnants of what once were larger, more continuous, woodlands. One critical factor appears to be the intrusions of predators, such as raccoons or skunks or house cats. These predators live mainly in the surrounding agricultural or suburban landscapes, but they make forays into the remnant patches and take their toll on the birds living in them. Think, for example, of the number of house cats in an average suburban neighborhood, nearly all of them subsidized by having predictable meals twice a day, but most also making nightly hunting expeditions into the woods across the street. Even if a bird attempted to nest in such a place, it would be unlikely to succeed. The absence or scarcity of many bird species from such places strongly suggests that they know, in an evolutionary sense, the futility of attempting to breed there.

Another important principle of landscape ecology is the concept of population sinks and sources. Within the larger ranges of most species, there will be habitat patches where conditions usually result in successful reproduction, often so successful that more young are produced than can be supported there over the long run. These places are called "source areas," because of the production of surplus young. Elsewhere within a species' range there will be other habitat patches where individuals may be able to survive but where they can reproduce little if at all. Such marginal places are called "sink areas," and a species may persist there only because of continuous immigration of individuals dispersing out of nearby sources. Key differences between sink and source habitat patches might include food availability, nesting sites, and numbers of predators.

Long-term conservation of many species obviously depends on our ability to distinguish sinks from sources. This in turn involves more than just counting birds; it also requires us to measure reproductive output and survival in different habitat patches. This is a difficult but important business, and one in which many ecologists and conservation biologists are actively engaged.

In some ways, small remnant patches of woodlands or grasslands in the Midwest or Northeast appear to function like islands, supporting fewer and often different kinds of species than similar but larger habitats on the "mainland."

But wasn't sycamore #70 also like an island? There it was, all alone in the sacaton grasslands of Lower O'Donnell Canyon, and yet just as rich in bird life—mostly the same *kinds* of bird life—as the forest of sycamores fifteen miles away in Lower Garden Canyon.

We do not know if the orioles and kestrels and kingbirds nesting in sycamore #70 produced as many young as those nesting in the sycamores over in Garden Canyon. That would be a very useful thing for somebody to find out. However, the presence of so many birds in that one tree tells us that a lot of birds at least *thought* it was a good place to be. Were the principles of landscape ecology not at work here? Were the birds not able to distinguish a sink from a source? If not, why not?

The answer to these questions may lie in a fundamental difference between landscape patterns in eastern versus western North America, and it may call for a modified "western" version of landscape ecology. One need not argue for vast, uninterrupted expanses of pre-Columbian forests across the eastern half of the continent to make the obvious point that western landscapes probably always have been more spatially heterogeneous than their eastern counterparts. Topographical variety alone is sufficient to make the case. The West has always been a much more complicated place than the East.

The arena in which many western landbirds evolved would have been a naturally fragmented landscape. This is not absolutely true of all western species everywhere, but it almost certainly would be the case for birds attempting to breed in most southwestern riparian woodlands. Even the largest stands in our study were narrow linear strips running parallel to the drainages they followed, and they probably always have been. Birds seeking a forest interior—a refuge from intruders from another landscape—would have found no place to hide. They had to arrive at alternative and perhaps quite varied solutions to self-preservation.

It is tempting to conclude that birds in riparian and other habitats in the Sonoita Valley enjoy an advantage over their midwestern or eastern counterparts, having evolved to tolerate or perhaps even require heterogeneous landscapes. They might, moreover, be preadapted to further fragmentation of their favorite habitats due to human activities such as livestock grazing, tree cutting, and depletions of the water table. Sycamore #70 may once have been part of a continuous strand of riparian trees along Lower O'Donnell Canyon, most of which failed to survive changes in the watershed at the end of the last century. Yet sycamore #70 persists, and the results of our study suggest that one sycamore is vastly better than none.

To a certain degree this gives us reason for optimism about riparian birds and other wildlife in the Sonoita Valley. The animals may not mind, or might even prefer, a fragmented landscape. But this is a risky assumption, because another kind of habitat fragmentation is working its way east across the Sonoita Valley, and the birds may not be prepared for this one at all.

The town of Sonoita is not an official incorporated place, but just a collection of houses and small service businesses clustered at the west end of the valley, with its back up against the foothills of the Santa Rita Mountains (Map 1). When we first started working in Arizona in 1974, Sonoita was close to being a proverbial wide spot in the road. There was a place to get gas and basic groceries, two restaurants (good ones, in fact), and some houses, most of which were on big lots and mercifully tucked out of sight in the oaks south of town.

A night trip across the Sonoita Valley used to be a drive in the dark. That is no longer the case. Like most places in the sun belt, Sonoita has begun to grow, spreading eastward across the open grasslands toward Bald Hill (Fig. 25). Leading the way are little outlying knots of homes, now mere islands in the sea of grass. But that is bound to change. There is enough private land between Elgin and Sonoita for lots more development, and we would be fools not to anticipate it.

FIGURE 25. Suburban Sonoita in 1998. (Photo by Bill Branan)

When the Sonoita Valley becomes filled with refugees from Sierra Vista and Tucson, or for that matter from Los Angeles and Boston, how will its wild things respond? Will the houses and fences and one-horse pastures and barns and small vineyards be just another kind of fragmentation that the birds can take in stride? Or will things be different?

We do not know the answers to these questions about the Sonoita Valley. No one has yet quantified bird numbers or reproductive success in those parts of the area that already are "settled," to apply an old term you don't hear much anymore. Somebody ought to do this, using the Research Ranch as a control. In the meantime, there is every reason to err on the side of caution, particularly if you are an individual or a planning board or a citizens' group concerned with long-term land use in the Sonoita Valley.

Evidence coming in from elsewhere in North America suggests that human occupation can have negative effects on wildlife even under what appear to be the most benign of circumstances. Focusing on the West, it can happen even in places with high natural landscape heterogeneity, where housing density is low or clustered in between open spaces, where there is little or no direct persecution of wildlife, and where the majority of the land remains as it was before development.

Ecologists and conservation biologists are only just beginning to understand what is going on, and we are reluctant to do more than point to avenues for further research. Among the possible causes of wildlife declines in suburbanizing landscapes are direct interference at nests and roost sites by people and their pets, the impacts of predators (especially house cats), and bird feeders.

Bird feeders? Some of the most enthusiastic consumers at backyard feeders, it turns out, are birds in the family Corvidae (jays, crows, and their relatives). A genuine gourmet item for a jay is the egg or nestling of another kind of bird. One of our graduate students, David Craig, found that jays were much more abundant in a low-density suburbanized forest in Colorado than in a comparable area without houses, probably because of bird feeders that helped them survive the winter. Then, in spring, this increased number of jays went out and searched for bird nests. The result was that avian reproductive success among other birds was much lower in the suburbanized area than it was in the undeveloped forest. It is a humbling thought that something apparently beneficial to wildlife, such as bird feeders, potentially could turn a source into a sink.

The issue of suburban sprawl in the West is one of the least tractable environmental issues to confront, because it is hard to find the enemy. Where there is a greedy and polluting industry at work, or a government agency that is not willing to do its job, we can point fingers and demand change. This is not to belittle the accomplishments of environmentalists who have taken on such causes; sometimes it takes great courage to point accusatory fingers, especially at somebody bigger and stronger than you. But it is even tougher to point your finger at an enemy out there, only to discover that the finger has turned around and is pointing back.

In terms of suburban expansion, we are *all* the bad guys. The Sonoita Valley is a spectacular and inspiring place to live. Having moved to such a place, there is a powerful urge to pull up the drawbridge and stock the moat with alligators. But there is no justice in that, no humanity, no equity, and not even much of a righteous cause. Except this: at some point the Sonoita Valley could be loved to death, and then it will be gone for everyone. No matter how virtuous we all might become as individuals, a sufficient number of us doing perfectly ordinary and moral things will become a force of destruction. It probably would be all right to use up a place like the Sonoita Valley, as long as there was always another one just down the road. For much of human history that has been the case, or so it seemed. But no longer.

Sycamore #70 remains a safe haven for birds into the foreseeable future, thanks to the Appletons and the Whittell Foundation and the National Audubon Society. The same cannot be said for most of the rest of the Sonoita Valley, and for lots of other good places, until we finally come to grips with the realities imposed by a world that isn't getting any bigger.

Ten

THE NEST BOX EXPERIMENT

One November afternoon in 1986 we were having coffee in our backyard at East Corrals, looking off down O'Donnell Canyon toward sycamore #70. It was one of those late fall days when much of North America is starting to worry about winter, but when things are just about perfect in southeastern Arizona. The leaves had fallen from the branches of sycamore #70, revealing the white and silver bark normally hidden under its canopy. Sacaton and other grasses on the slopes of the canyon were cured to their own particular shades of gold and russet and bronze. The sky was clear blue, the air was still, and there was just enough bite in the air to make sitting in the sun a pleasure.

We already had three years' data in the bank on the riparian bird project. Some interesting patterns had become clear, and Tom Strong was busy writing his doctoral dissertation on these results. But standing in the shade counting riparian birds makes an invigorating counterpoint to studying grasslands, especially in June; and so our talk that afternoon turned to possible ways we might profitably continue the study.

The conversation went something like this:

If the science of ecology has one central purpose, it is to discover why species are common in some places but missing or rare in others. As it happens, two very different views have emerged over the years on this topic. One holds that most species are distributed independently of one another, according to their own individual habitat requirements. Another view is that competitive interactions between species can be very strong, so that the presence of one negatively impacts the abundance of another. According to the first view, groups of species living together are just chance assemblages of individuals that happen to share resource needs. According to the opposing theory, "biological communities" are predictably structured assemblages whose member species can coexist because they are sufficiently different to tolerate one another's presence.

We can illustrate the implications of these two theories with a simplistic example. All woodpeckers nest in cavities, so hollow tree trunks are a critical resource for each species. Woodlands have lots of different kinds of woodpeckers, while grasslands have none, because the first habitat is the only one providing the resources (trees) that woodpeckers need. According to the first (individualistic) view of ecology, this is about all that can be said about woodpecker communities. Not all forests have all kinds of woodpeckers, of course, but that may be because a particular woodland happens to lack a resource—a certain kind of insect food, say—that a specific woodpecker requires, or because the woodland grows in a climate that a particular woodpecker cannot tolerate. According to the second (community) view, the woodpeckers that coexist in a woodland are there not only because it provides what they need but also—and critically—because they are the long-term winners of a competition with other kinds of woodpeckers for those resources.

The individualistic versus community issue persists in ecology, not only because the evidence is equivocal, but also because the two theories appeal to different sorts of ecologists. In the first case, nature looks substantially chaotic. Each individual species does its own thing, takes its own ecological chances, and stands or falls according to its own evolutionary race against changing environmental circumstances. According to adherents of the individualistic theory, biological communities are largely artificial constructs that exist mainly in the eyes and minds of ecologists with a tidiness fetish.

To community ecologists, nature looks relatively predictable and organized, since it is put together according to rules. Among other things, these rules dictate that no two species with very similar ecological requirements can coexist indefinitely, because one eventually will outcompete the other. Adherents suggest that if we can discover the rules of community ecology, we can formulate principles that are as solid and fundamental as the laws of matter and energy. Skeptics have called this approach physics envy.

Like most great debates in ecology, the individual versus community controversy appears to be in no danger of resolving itself. Probably there is some truth to both views, and that is part of the problem. Another is that it can be difficult to gather evidence allowing a rigorous test of the alternatives. For birds, there is good evidence that some coexisting species share most critical resources (contrary to predictions of the community theory) and that their collective numbers are driven up and down in synchrony, not by competition but largely by unpredictable impacts of the weather. By contrast, bird species belonging to certain other assemblages often have remarkably complementary ecologies (as predicted by the community theory), with each species occupying an ecological niche distinctively different from those of all others in the group.

The problem is that the evidence usually is observational, not experimental, and therefore circumstantial. We are describing patterns in nature, often carefully and clearly, but we are not testing for the causes behind those patterns. If there is one area of agreement among ecologists, it is about the power of conducting ecological experiments in nature as a means of looking for causes behind patterns. Suppose we are interested to learn if woodpecker A is excluding woodpecker B from a particular place, or at least holding down its numbers. We could remove most or all individuals of species A, and then see if species B moves in or becomes more abundant. Like all good experiments, this one would need a control—some place or places where species A is not removed. Here, we could learn whether species B is increasing its abundance anyway, thanks to circumstances not related to the experiment.

While birds have played a prominent role in studies describing patterns in ecological communities (or the lack of them), they have proven to be very difficult subjects for field ecological experiments. Removing or reducing populations of potentially competing birds, at a scale meaningful to the birds, usually is a wrong thing to do. Certainly it is not appropriate for the Research Ranch.

This brings us back to the riparian birds in the Sonoita Valley. As we were watching sycamore #70 on that fall day back in 1986, it occurred to us that there might be a way to test the individualistic versus community theories in an experiment that both was appropriate and that took advantage of our existing data base about riparian birds.

We had already documented the relative abundances of more than seventy species on 132 plots. Although the ecology of each species was individually distinctive, each could be placed into one of two categories: seventeen of the species nested in cavities in trees, while the remaining fifty-three species built cup nests out in the open, on tree branches, in shrubs, or on the ground. An extensive literature shows that local densities of cavity-nesting birds can usually be increased by hanging artificial nest boxes on tree trunks, because the supply of natural cavities often is limited. This is particularly true for so-called secondary cavity-nesters, which cannot build their own nests but must use old ones abandoned by primary cavity-nesters, mostly woodpeckers.

We were confident that we could increase the densities of secondary cavity-nesters on our riparian bird plots by adding nest boxes. But the main point of such an experiment would be something else. Suppose the entire avian assemblage were limited by other shared resources, such as insect foods. If this were the case, we predicted that densities of open-cup nesters should decline on plots where the artificial nest boxes had been added and where densities of cavity nesters had increased. Such a result would support the community theory of bird distribution and abundance, because it would suggest that competition between species was an important factor determining the composition of the entire avian assemblage. If, however, increased densities of cavity nesters had no impact on densities of open-cup nesters, this would support the individualistic theory of avian distribution and abundance, because it would suggest that changes in abundance of one group had no impact on abundance of the other.

The field experiment we designed and implemented was doubly controlled, in the following way. We had three years' (1984–1986) data on the relative abundances of birds on all the riparian plots before any manipulations (nest box additions) were performed. In the spring of 1987

FIGURE 26. A bluebird delivering food to its young at an artificial nest box. (Photo by William Ervin)

we added three nest boxes each to half the plots, but we left the other half as permanent unmanipulated controls. Then two students, Chuck Aid and Alex Cruz Jr., continued counting birds on all the plots for another three years (1987–1989). The resulting data allowed us to ask two questions. First, did densities of cavity and/or open-cup nesters change on the experimental plots after we added nest boxes to them? Second, did densities on the control plots remain more or less the same over the entire six-year period?

Six species used our nest boxes over the three-year experimental phase of the study (Fig. 26). In decreasing order of enthusiasm for the artificial nests, these were Bewick's wren, bridled titmouse, ash-throated flycatcher, eastern bluebird, white-breasted nuthatch, and dusky-capped flycatcher. Their combined use of the boxes was sufficient to nearly double our counts of all cavity nesters combined on the experimental plots, compared to what they had been before the boxes were added. On the control plots, in contrast, relative abundances of cavity nesters changed hardly at all over the six-year period.

So we succeeded in our first goal, to dramatically increase local densities of the cavity-nesting bird community on the experimental plots. But was there any response on the part of the open-cup nesters? Here things got more complicated, and provided a lesson in the importance of controls. As it turned out, the total assemblage of open-cup nesting species actually increased in abundance on the experimental plots after the nest boxes were added, but only by about 7 percent, whereas the increase on the control plots over the same period was 33 percent (Fig. 27). In other words, for entirely separate reasons lots of open-cup nesters were on their way up between 1984 and 1989, and nest box additions significantly retarded that growth.

The ten open-cup nesting species responding most negatively to our experiment were the American robin, western wood-pewee, common yellowthroat, blue grosbeak, eastern meadowlark, yellow warbler, Grace's warbler, hepatic tanager, rock wren, and house finch. This represents a diverse ecological array of species, not matched in any obvious ways with the six cavity-dwellers that used our nest boxes. Competition between cavity and open-cup nesting species therefore appeared to be communitywide, rather than involving specific one-on-one interactions between particular pairs of species. Ecologists term this sort of relationship diffuse competition; it has been demonstrated in the field rather infrequently.

This particular field experiment will not resolve the community versus individualistic debate. The results do, however, suggest some level of community organization among the great variety of birds nesting in riparian habitats in southeastern Arizona, insofar as the whole assemblage appeared to be limited by competition for shared resources.

A number of nest box experiments have been performed elsewhere, and for the most part they have failed to detect any evidence for com-

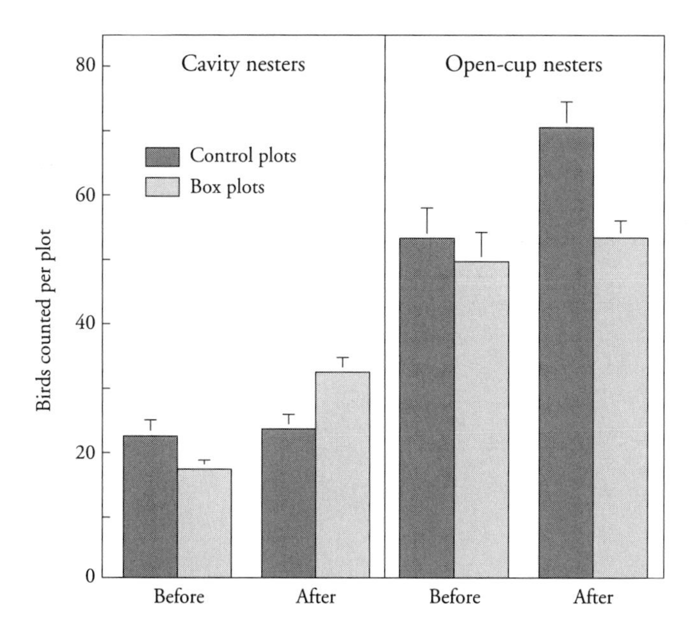

FIGURE 27. Numbers of birds counted on experimental box plots before (1984–1986) and after (1987–1989) addition of artificial nest boxes, and on unmanipulated control plots. Bars are means with standard errors. (Data from C. E. Bock et al. 1992)

petition between open and cavity-nesting species. These other studies, including one in the pine forests of northern Arizona and one we conducted in the Colorado Front Range, have involved high-altitude or high-latitude forests with lower bird densities and relatively few species. In these places, interspecific competition may have been much less important than climatic variability in limiting populations of the relatively few kinds of birds present. By contrast, southwestern riparian habitats support some of the highest densities of birds ever recorded, and it is in such highly productive and diverse ecosystems that insect-eating birds are most likely to limit their prey, and presumably therefore to be involved in intense competitive battles over limited food resources.

The community versus individualistic theories, unlike the laws of physics, may have an underlying geography. There also is a conservation lesson to be learned from our nest box experiment, which has to do with ecological space. At least in southern Arizona, enhancing one group of species may be accomplished only at the expense of another. At the end of the 1989 field season we took down most of the nest boxes we had hung up in the Huachuca Mountains and across the Sonoita Valley. It seemed the right thing to do. Let the cavity- and cup-nesting birds work things out for themselves.

Eleven

PLAINS LOVEGRASS

Plains lovegrass (*Eragrostis intermedia*) is a tall bunchgrass that has become much more common on the Research Ranch since livestock were removed in 1968. It starts growing in early summer, before the grama grasses, and by late summer its characteristic branching seed heads can give whole hillsides a striking purple hue. In fall, plains lovegrass seed stalks break off from the parent plants and tumble across the landscape. Frequently they catch and accumulate in the branches of oaks, mesquites, and shrubs, giving these woody plants a shaggy appearance that lasts through the winter.

To those of us working and living on the Ranch, plains lovegrass became a living symbol for the sanctuary as a whole, and for its overall purposes (Fig. 28). As things turned out, it also provided a powerful lesson in the value of long-term ecological studies, and the dangers of doing anything less.

As we noted earlier, drought, fire, and grazing are the most important environmental forces at work in all the world's grasslands. These factors consequently drive the adaptation and evolution of most grassland plants and animals. Because each of these three forces is inherently episodic, grassland ecosystems are highly variable across both space and time. Droughts come and go. Lightning strikes happen to

FIGURE 28. A mixed stand of blue grama and plains lovegrass on the East Mesa in 1988, one year after the Big Fire. Note burned mesquite trees. (Photo by the authors)

start fires in some times and places but not in others. Native grazing mammals usually travel in herds, and they frequently are nomadic both seasonally and annually. In North America, the three forces may have come together only recently, as most of this continent's grasslands appear to be relatively young compared to other sorts of ecosystems.

The three forces also are strongly interactive. For example, droughts can increase the likelihood of fire by reducing the moisture content of accumulated fuels. However, in heavily grazed or recently burned areas, a drought can reduce plant productivity to the point where there is nothing left to burn at the end of the growing season. The absence of grazing can result in the buildup of dead plant material—burnable fuels—which increases the probability, intensity, and size of wildfires. Because fires can increase the subsequent palatability and productivity of grasses, herds of grazing mammals are likely to be attracted to recently burned areas. Grazing by fenced, predator-proofed, and variously subsidized domestic livestock can have very different ecological effects than grazing by their free-ranging wild relatives, because livestock usually cannot be nomadic or opportunistic in relation to droughts and fires.

No organism on the Research Ranch has demonstrated the interactive importance of drought, fire, and grazing more dramatically than plains lovegrass. Through most of the 1980s we and other researchers found it to be consistently more common inside than outside boundary fences of the sanctuary. We also had evidence of its increase on the sanctuary between the late 1960s and the mid-1980s. It seemed obvious that this species responded negatively to grazing, or at least positively to livestock removal. Other grassland ecologists had reached the same conclusion elsewhere in the Southwest. We anticipated a continuing steady increase of this bunchgrass on the Ranch in the absence of cattle and other disturbances, the eventual result being some new equilibrium in which plains lovegrass would be the dominant grass on most parts of the sanctuary.

In the summer of 1989, however, plains lovegrass, our living symbol of the Research Ranch, began to die. The monsoon was late and very weak that year (see Fig. 6), and although by August the other common bunchgrasses were almost as lush and green as usual, scattered among them were conspicuous gray-brown patches of plains lovegrass—mostly clumps of dead leaves left over from the previous year.

We did not quantify plains lovegrass mortality in 1989, and we made no systematic effort to survey all of the sanctuary or surrounding lands. Qualitatively, however, three very intriguing patterns became apparent as the summer wore on. First, we saw no sign of mortality among the plants in grazed pastures surrounding the Ranch. Second, stands of plains lovegrass in different parts of the sanctuary showed very different amounts of mortality. Some mesas and hillsides were little affected, while on others nearly all the plants were dead. Most places where plains lovegrass survived best were areas that had burned in wildfires within the past three years. And finally, even in areas where mortality was widespread, one place where the plants were healthy and green was along the edges of roads.

Over the winter of 1989–1990 we developed three hypotheses about the population dynamics of plains lovegrass, based on our impressions from the previous summer:

1. Plains lovegrass plants are vulnerable to drought.
   Evidence: 1989 was a dry year; plants survived better along roadsides, where runoff water collects.
2. Frequent fire reduces plains lovegrass vulnerability to drought.
   Evidence: there was little or no sign of mortality in areas that had burned recently.
3. Grazing reduces plains lovegrass vulnerability to drought.
   Evidence: there was little or no sign of mortality in grazed areas surrounding the sanctuary.

Over the next three summers we collected and compiled data to test these three hypotheses.

In the summer of 1990 we measured the densities of live versus dead plains lovegrass plants under various circumstances on and adjacent to the sanctuary (Table 4). These data indicate that mortality was highest in ungrazed and unburned habitat and lowest in grazed or recently burned habitat. Despite the absence of mortality in the grazed areas in 1989 and 1990, density of living plains lovegrass still was lower there than in ungrazed areas, regardless of fire history.

The loss of plains lovegrass from ungrazed and unburned stands was alarming. At the end of the 1990 field season we wondered if the mortality might continue until the species disappeared from unburned areas of the sanctuary. Certainly our model projecting a steady post-grazing increase of this grass was in jeopardy.

One of the advantages of long-term research is that all sorts of information begins to accumulate that turns out to be useful in unpredictable ways later on. Such was the case with our work on plains lovegrass. As part of another study we had collected data on grass canopy cover inside a series of twenty-five circular plots, each two hundred

TABLE 4. Densities of living and dead plains lovegrass in 1990 on and adjacent to the Research Ranch

	Number of plants per 10 m²	
	Alive	*Dead*
Grazed and unburned	39	0
Ungrazed and unburned since 1968	46	105
Ungrazed and burned within 1–3 years	133	2
Ungrazed and burned within 4–6 years	80	62

square meters in area, in 1984, 1987, 1988, and 1989. Fourteen of these plots were on the North Mesa, and these had remained unburned at least since the Ranch was established in 1968. The remaining eleven plots were on the East Mesa, and these had burned in the Big Fire of July 17, 1987 (see Chapter 6). We sampled these plots again in August of 1990, 1992, and 1993, to generate a long-term picture of plains lovegrass canopy cover before and after the 1987 fire on the East Mesa, with the unburned North Mesa as a control site.

Plains lovegrass cover varied substantially between 1984 and 1993, in ways that suggested the importance of both precipitation and fire for maintaining its vitality (Figs. 29 and 30). Although mortality on the unburned plots in 1989 was associated with a relatively dry year, plants nevertheless continued to die through the summer of 1990 despite above-average precipitation. By 1993, however, plains lovegrass cover had increased on the unburned plots to higher levels than before the 1989 drought.

Recent fire, we concluded, had buffered plains lovegrass plants against the effects of the 1989 drought. Although there was a temporary reduction in plains lovegrass cover immediately after the fire in 1987, few if any plants were killed, and nearly all survived the 1989 drought. By 1993, plains lovegrass canopy cover on the previously burned plots was more than twice that on the unburned control area.

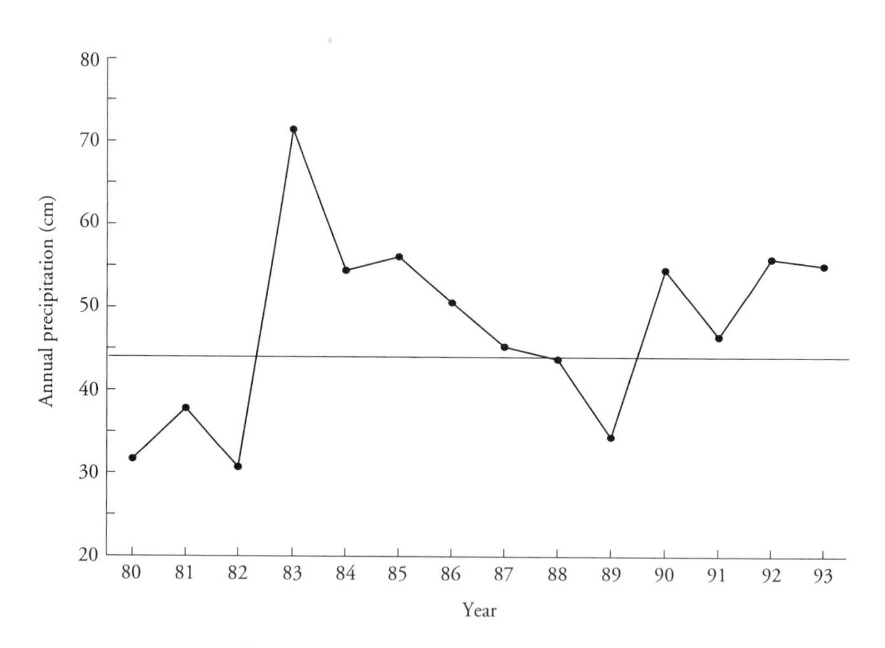

FIGURE 29. Annual precipitation at the Research Ranch, 1980–1993, calculated as rainfall for that year's summer growing season (July–September) plus precipitation over the preceding nine months (October–June). The horizontal line is the twenty-year mean. (Redrawn from C. E. Bock et al. 1995)

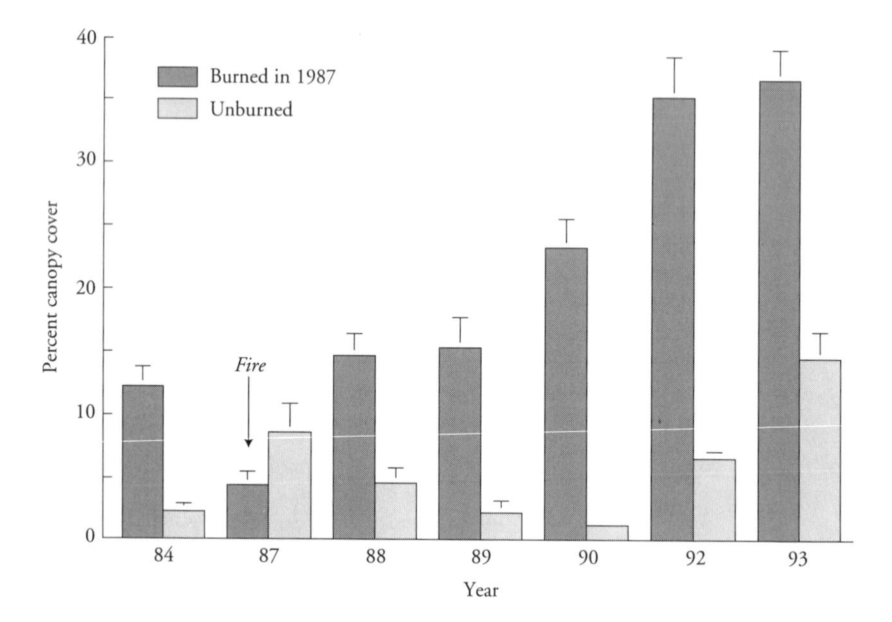

FIGURE 30. Percent canopy cover of plains lovegrass on fourteen 200-m^2 plots unburned from 1984 to 1993 and on eleven plots burned in July 1987. Data were collected at the end of the growing season, in August of each year. Bars are means with standard errors. (Redrawn from C. E. Bock et al. 1995)

Based on these findings, we believe the following things to be true about plains lovegrass population dynamics on the Sonoita Plain:

1. Substantial mortality can occur during periods of below-average precipitation. There also may be a residual effect, with mortality continuing past the end of a drought for at least one growing season.
2. Recent fires in some way protect plants against the effects of droughts, such that the highest densities are most likely to occur in areas that have burned within the past three years. Winter accumulations of seed stalks probably increase the likelihood of fire, thus ensuring continued vitality.
3. Livestock grazing also reduces drought-caused mortality, but densities of plains lovegrass in grazed areas usually are lower than those in ungrazed areas. A possible factor could be reduced seedling recruitment in grazed habitat.
4. Even in the absence of fire, plains lovegrass will recover from a drought in ungrazed areas once precipitation has risen to some critical level.

Much remains to be learned about the biology of plains lovegrass. For example, we do not know how burning or grazing might actually protect the plants against the effects of drought. Also, we know nothing about seed production and seedling establishment, or how these vary in relation to precipitation, fire, or grazing. Despite these uncertainties, one thing is abundantly clear: the dynamics of plains lovegrass are much more complex than we originally suspected. No evidence suggests that it is increasing steadily toward any sort of long-term population stability, on the sanctuary or anywhere else.

Classical models of plant ecology describe how a biotic community will recover from a perturbation by moving through a predictable series of successional stages, until eventually the vegetation returns to the predisturbance condition. For example, a fallow field in the midwestern United States will be colonized first by native annual plants, sometimes called weeds, which in turn are succeeded by perennial herbs and shrubs, which are then followed by a pine woodland, and eventually by a hardwood forest like the one that was there before the field was cleared for agriculture in the first place. This final and relatively persistent stage is called a climax community, and it used to be generally accepted that most of the terrestrial world would exist in such a state if we could just get humans out of the picture. This notion in turn led to conservation strategies designed to preserve what was left of the climax vegetation in a particular area, and to attempt restoration of the rest.

Ecologists now recognize that most of the natural world is inherently variable, with or without humans. Fires, floods, droughts, landslides, hurricanes, avalanches, and other physical disturbances continually and naturally operate to set the successional clocks back to zero. Furthermore, humans have been around for so long by now that we must ask whether a landscape without them is natural or not. This is particularly true of grasslands, because we are the world's preeminent grassland animal.

Does this mean that grassland conservation is a waste of time, or an unnecessary intrusion upon the rights of others wishing to use them for personal gain? The answer, absolutely and emphatically, is no. Advocates of the so-called wise use of our natural resources belittle conservationists by accusing them of trying to preserve a world that never existed. This is an erroneous accusation, and they ought to know it.

The biological diversity of our planet exists for many reasons, but perhaps the most important one has to do with environmental heterogeneity. The world is both spatially and temporally variable, and it is this inherent variability that has provided the world's flora and fauna with most of their ecological and evolutionary opportunities. Human activities threaten biological diversity not so much because they disturb the environment, but because they homogenize it. For example, humans have been clearing tropical rain forests for agricultural purposes for millennia. The threat to these forests today is not that we want to cut them down; it is that we want to cut them all down at the same time. We wish to disturb these forests on such a large spatial scale, over such a short temporal one, that few if any of them will be left in late successional or climax condition. Those thousands of species that do best in recently cleared forests will thrive, but those other thousands of species that depend on a mature forest are doomed.

And so it is with the world's grasslands. Droughts, fires, herds of grazing mammals, and their predators, human and otherwise—in the old days all of these could come and go with their own independent and interactive rhythms, with none operating to the exclusion of the others. The result was a landscape mosaic, offering abundant opportunities for those species requiring wet or dry, burned or unburned, grazed or ungrazed. Now, though, we humans are such a numerically dominant force in most grasslands that we are driving them toward whatever state of environmental uniformity best suits our needs.

The Appleton-Whittell Research Ranch is not a sanctuary because it harbors plants and animals with greater inherent value than those living on any other 7,800-acre piece of the Sonoita Valley. Plains lovegrass is no more to be cherished than blue grama. The bunchgrass lizard commands no more respect than any other reptile, nor the Botteri's sparrow more than any other bird. But given the virtual ubiquity of livestock across the southwestern plains, the Ranch is a sanctuary because it is an unusual place in an unusual condition, and it provides opportunities for plants and animals that otherwise would have few places left to live.

Twelve

COTTON RATS AND REAL DOGS

RODENTS, THOUGH MUCH less influential than cattle or people, are the most abundant mammals on the Sonoita Plain. Yet because wild mice and their relatives mostly are nocturnal and inconspicuous, it is necessary to set traps to find out who is out there, in what numbers, and in which places.

The simplest metric by which mammalogists compare rodent abundances across time or space is in numbers captured per trap-night. One trap-night involves a live-trap being baited with something rodents like, such as rolled oats or peanut butter, and placed out for one night. Traps are opened in the evening and then closed the following morning, before it gets too hot, and after the rodents have been identified to species and released. We counted up the other day and discovered that our studies of rodents on the Ranch have involved over twenty thousand trap-nights. That's a lot of peanut butter.

We began live-trapping rodents on the North Mesa of the Research Ranch in the summer of 1981, as part of an overall effort to compare the ecology of grazed versus ungrazed grasslands. The commonest rodent we caught was the deer mouse, one of North America's most generalized and widespread animals—so much so that a mammalogist colleague, Dave Armstrong, has dubbed it our National Mouse. Our data

FIGURE 31. A cotton rat. (Photo by the authors)

for 1981 and 1982 fully supported that appellation, not only because deer mice were abundant, but also because they were about equally common on both sides of the fence.

Other kinds of rodents, in contrast, differed dramatically across the fence line separating grazed from ungrazed pastures. Merriam's kangaroo rat is a species that prefers open areas with exposed sandy soil; it was eighteen times more common in grazed habitat on the North Mesa. All of the other rodents we captured were species typically associated with grasslands, and they all were substantially more abundant on the ungrazed side of the fence. Most notable among these was the tawny-bellied cotton rat (Fig. 31). Cotton rats are grazing rodents, usually restricted to areas of heavy grass cover, and so it is not surprising they were much more common on the sanctuary than across the boundary fences.

Different research projects brought us back to the North Mesa in 1984–1985 and again in 1988–1990, but unfortunately only to the ungrazed side of the fence. Recently we assembled and compared rodent trapping data for the whole ten-year period. The results were startling, and they suggest that our earlier picture of rodent population dynamics in the absence of livestock was far from complete.

The trapping circumstances in 1984–1985 were essentially the same as three and four years earlier, except we had only one line of traps crossing the ungrazed mesa top. From 1988 to 1990 we followed an entirely different approach. We had built a series of sixteen cages on the mesa, the original purpose of which was to exclude insect-eating birds and then to allow comparison of grasshopper densities inside and outside the cages (see Chapter 8). Our songbird cages also necessarily excluded larger predators such as rattlesnakes, hawks, owls, and carnivorous mammals that feed mainly on rodents. Incidental to the songbird/grasshopper project, therefore, we regularly set rodent livetraps in the cages and on adjacent open plots, to determine if predator protection also was having an effect on mice and rats.

These data, combined with the results of our earlier trapping efforts, suggested two interesting and important things about rodent populations on the North Mesa (Fig. 32). First, numbers of most species, and especially deer mice and cotton rats, declined dramatically between 1981 and 1990. Second, total numbers of rodents captured, and especially numbers of cotton rats (but not deer mice), were much higher inside than outside the predator exclosure cages.

Taken together, these results strongly suggest that at least some of the long-term rodent declines on the North Mesa could have been due to increasing predation pressure. Cotton rats in particular were noticeably more abundant inside the cages. Unfortunately, because we failed to continue trapping on the grazed side of the fence, we do not know if the declines occurred there as well, or what effects predator exclusion might have had on rodents in grazed as well as ungrazed habitat.

Deer mice virtually disappeared off the North Mesa between 1985 and 1988: we caught only three in over six thousand trap-nights during 1988–1990. This was a surprising result in light of the normal abundance of this generalized rodent in most grasslands across the continent. The decline could have been due to increased predation; if so, our cages did not provide sufficient refuge for deer mice to make a

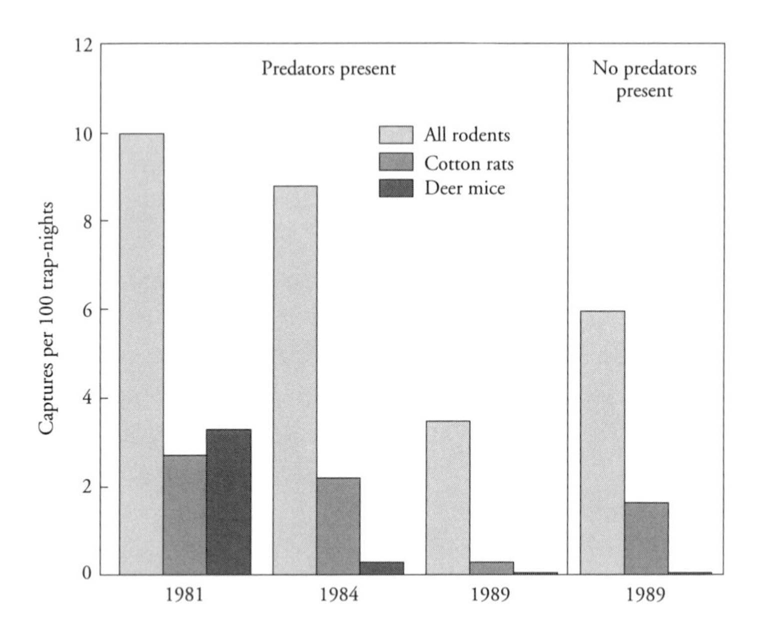

FIGURE 32. Relative abundance of cotton rats, deer mice, and all rodents combined in different years on the North Mesa of the Research Ranch. In 1989, rodent numbers were compared between cages from which predators were excluded (see Fig. 22) and open grasslands where predators were present. (Data from C. E. Bock at al. 1984; and C. E. Bock and J. H. Bock 1994)

comeback. Another possibility is that post-grazing vegetation changes had rendered the ungrazed part of the North Mesa somehow unsuitable, even for the National Mouse.

Did predators become more common on the Research Ranch between 1981 and 1990? It seems likely, but we do not know.

Rodent predators that hunted the North Mesa in those years included Mojave rattlesnakes, red-tailed hawks, northern harriers, great horned owls, bobcats, badgers, and coyotes. We have circumstantial evidence that at least one of these, the coyote, may indeed have played a role in reducing rodent numbers.

Few animals provoke such strong and varied opinions as coyotes. They are "ruthless" if they kill livestock, and they are "useful" if they kill mice and jackrabbits that otherwise might "overrun the range."

Despite relentless and widespread persecution, coyotes remain abundant in the West. This is attributable partly to their adaptability, and partly to the extermination of wolves, which normally dominate coyotes but whose lifestyles made them much more vulnerable to the obscenities of predator control. Given the long-overdue comeback of wolves in the West, coyotes had better keep an eye out.

Coyotes, of course, are neither ruthless nor useful. They just *are,* and at the Research Ranch they were welcomed as a part of the natural scene. If not more abundant, they were at least more conspicuous and less wary here than many other places in the West. We found dens on the sanctuary where coyotes raised their pups, some surprisingly close to roads and houses. Coyotes hunted the mesas and washes at dawn and at dusk, mostly alone but sometimes in pairs. They watched us as intently as we watched them, and usually they did not run unless we got too close. Our coyotes howled at night, and sometimes during the day as well, especially when they had pups. Local folklore has it that a coyote howling in the daytime means it is going to rain. Perhaps this is true.

Thanks to the efforts of Vern Hawthorne and Joseph Ortega, we learned something about how coyotes made their living on and near the sanctuary. Vern was Ranch manager from 1980 to 1983. Between May 1982 and May 1983 he methodically collected 759 scats deposited on the sanctuary by the local coyote population. It fell to Joe, one of our graduate students, to take the scats apart and find out what all those coyotes had been eating.

Research Ranch coyote scats, Joe found, contained the digested remains of desert cottontails, black-tailed jackrabbits, rock squirrels, pocket mice, woodrats, harvest mice, cotton rats, mule deer, white-tailed deer, pronghorn, peccaries, cattle, grasshoppers, beetles, moths, spiders, sparrows, lizards, grass, acorns, juniper seeds, cactus fruits, and mesquite beans. Some of the meat doubtless was scavenged from animals that were already dead, including cows. Lest we denigrate the coyote's feeding habits, remember that humans also like to gather occasionally for a dead-cow feast.

Fur and bones of cotton rats occurred in nearly one-third of the 759 scats, making these rodents second only to rabbits as the most popular item on the local coyote menu. Without doubt, coyotes hunting the North Mesa killed a lot of cotton rats, though some of these and perhaps other rodents managed to find refuge in our cages.

Late one August night some coyotes came down the draw behind our house. Perhaps they were pups, out for the first time and learning to hunt cotton rats in the sacaton. We were unaware of their presence until they stopped just outside our open bedroom window and erupted into a great synchronized howl.

The wail of a coyote is like no other sound. Perhaps there is no such thing as genetic memory, but if it does exist, then a little piece of our DNA must remember a long-extinct grassland where our ancestors and the ancestors of coyotes once lived together and where that howl must have meant something important.

The feeling must be even more powerful for dogs. On that August night when the coyotes sang, Huachuca, our Labrador retriever, was sound asleep, curled up on her deluxe, extra-large L. L. Bean dog bed, which she used reluctantly when all the people beds were filled. The coyotes brought her instantly alert. Her back hairs bristled, her nose quivered, and her brow furrowed in a way that makes even a Labrador look intelligent.

We tried to read Huachuca's mind—an admittedly risky endeavor. She engaged in some desultory growling. Certainly a part of her reaction was aggressive and territorial. Yet there was more: a look in her eyes, an unmistakable contact, and a realization—*Those are the Real Dogs out there, doing Real Dog business.*

Now, Huachuca would never have willingly given up the amenities of her good life: a soft, dry place to sleep, a house that was cooled in summer and warmed in winter, and a predictable, twice-daily bowl of kibble. She marked the majority of her time in Arizona by lying around in the sun or the shade, getting scratched behind the ears, and eating. But she was most alive on those mornings we trapped rodents together on the North Mesa, or when she jumped a black-tailed jackrabbit up in Post Canyon, or even that time we surprised a Mojave rattlesnake out by Finley Tank.

It was much the same for all of us who worked and lived the better part of three decades on the high plains of southeastern Arizona. Daily immersion in the beauty and diversity and mystery of real-world ecosystems added a sense of connection to our lives that simply could not have been generated in any landscape dominated by humans.

Yet those grasslands also could be hot and dusty, or wet and sticky and full of chiggers and mosquitoes and no-see-ums. Worst of all, they never let us know when we finally got something right. At the end of a hard day, then, one of the best things of all was to come inside, have a shower and a drink, and talk things over. If we were lucky, an evening thunderstorm would roll down off the Mustang Mountains, and we could watch it from the comfort of a dry room with a good view.

We lived in touch with wild things on the Sonoita Plain, but we willingly wrapped ourselves in a twentieth-century security blanket the whole time. This may sound elitist and privileged, and perhaps it was. Yet the same opportunity could and should be available to anyone who wants it. Here is one of the greatest ironies: Just when technology and medicine finally have gotten to the point where human life can be mostly comfortable instead of mostly miserable, we crowd up so much of the planet that there is no room left to *be* comfortable in lots of the best places.

Thirteen

ALIENS

ONE DAY IN 1975 ARIEL APPLETON described for us a strange tree she had found growing below Post Canyon dam. Ariel is a keen observer of the natural scene and intimately familiar with the Research Ranch. This particular tree was like none she had ever seen before. It had rather wispy branches covered with gray-green scalelike leaves, and numerous pink flowers. It was growing with a stand of young cottonwoods that had become established along the Post Canyon stream bank subsequent to cattle removal in 1968.

The strange tree proved to be a tamarisk. Also known as saltcedar, this aggressive Old World plant was first introduced into North America in the nineteenth century as a backyard ornamental. Beginning in the early 1900s it escaped cultivation and began to colonize riverbanks and lakeshores throughout the lowland Southwest, often forming nearly impenetrable stands largely devoid of native plants and wildlife.

The tamarisk discovered that day by Ariel Appleton did not survive to see another sunrise. In this instance we chose to become manipulators rather than observers of nature, having seen wholesale destruction of riparian habitats by tamarisk elsewhere in Arizona and Colorado. This particular tree probably was a fluke, because the sanctuary is higher in elevation than most of the places where tamarisk thrives. But given the scourge it has become elsewhere, we took no chances.

Establishment and spread of exotic species has the potential to threaten the biological diversity and ecological integrity—indeed, the very purposes—of a place like the Research Ranch. Replacement of native flora and fauna with nonnative aliens would compromise its role as an ecological benchmark or reference point, as well as degrade its value as a haven for endemic species. Tamarisks have not recurred on the sanctuary, and there have been no other invasions of exotic trees or shrubs, and so its riparian woodlands remain intact and rich in native species. Unfortunately, the same cannot be said for all of its grasslands.

We once attended an international meeting of land and resource managers in Washington, D.C. At the end was the obligatory capstone banquet of overcooked roast beef and cold peas. We ended up seated at a table with a man who opened things up by stating: "You've heard of the Aswan Dam? Well, that's my baby!" He went on to explain the benefits of the dam to the Nile Valley, including flood control, irrigation, and other "improvements on Mother Nature."

"Improving on Mother Nature" is a phrase you don't hear much anymore. And a good thing too, suggesting, perhaps, that we have begun to replace arrogance with at least some humility about our position in the grand scheme of things. We cannot avoid manipulating natural landscapes for our own essential purposes, of course, but we ought to do it humbly and very carefully, given that so many of our past "improvements" have turned out to be otherwise, not only for the Earth but for ourselves as well. Nowhere is this more true than with the purposeful introduction of exotic species, and in the southwestern plains the premier example has to do with two grasses native to southern Africa.

Historic degradation of desert grasslands naturally led to attempts at restoration and revegetation. Because reseeding with native grasses had proven only marginally successful, land managers in the 1920s and 1930s began to search for exotic grasses that might be more tolerant of drought and perhaps easier to establish on rangelands already invaded by shrubs and trees. Exotic range and pasture grasses that had done well in the northern Great Plains and in the Great Basin, such as smooth brome and crested wheatgrass, were not adapted to southwestern heat and aridity. The search for new grasses persisted, eventually shifting to Old World environments similar to the American Southwest.

Range scientists finally hit pay dirt with the discovery, import, and seeding trials of a variety of African lovegrasses in the genus *Eragrostis* that were close relatives of the Arizona native, plains lovegrass. Two species in particular, Boer lovegrass and Lehmann lovegrass, established themselves and thrived in former desert grasslands where most native species were scarce or absent.

The introduction of these African exotics was heralded as a triumph of rangeland revegetation. Grasslands were restored, soils were stabilized, and some watersheds were protected. For decades, Lehmann lovegrass was the species of choice for reclaiming degraded southwestern rangelands and for reseeding bare soils along road cuts and at construction sites. It continues to be used for these purposes.

Planting the African exotics has proven to be a decidedly mixed blessing, for three reasons. First, these durable grasses not only established well where they were deliberately planted, they also spread into surrounding areas. Second, the African aliens formed such dense stands that they often excluded or retarded those native species that remained. Third, there was no obvious way to control their continued expansion. Once the genie was out of the bottle, there was no way to get it back in.

In retrospect, it is not surprising that the African lovegrasses proved so successful, and so invulnerable to subsequent efforts to control their spread. They were, after all, the result of a worldwide search. They thrive on fire, which is a part of the environment in which they evolved. They are prolific seed producers, as might be expected of any grass that is prized for its ability to establish quickly where it is intentionally planted. Whatever enemies they might have had—insects, parasites, diseases, and so forth—were conveniently left behind in Africa when the first seeds were brought to Arizona in 1932. They even provide relatively poor forage, so that livestock will eat out most of the native species before turning to the African exotics.

Many southwestern desert grasslands are now almost pure stands of exotic lovegrasses. A case in point is the Santa Rita Experimental Range, established in 1903 south of Tucson as a site for long-term studies of grassland ecology and range management. By the early 1990s, so much of the Santa Rita Range was dominated by Lehmann lovegrass that in many areas it would have been difficult to study the ecology of native desert grasslands at all.

Lehmann and Boer lovegrass apparently were first introduced to the Sonoita Valley by the U.S. Soil Conservation Service, when both species were planted on the Babocomari Ranch in 1940. Sometime during the next twenty years two areas destined to become parts of the Research Ranch also were seeded with the exotics, particularly with Boer lovegrass—one on the North Mesa and one on the East Mesa (Fig. 33).

When we first began our work at the Research Ranch in the mid-1970s, stands of exotic lovegrasses planted at least twenty years earlier remained intact, but they did not appear to be spreading into adjacent sites dominated by native species. In 1984 we were fortunate to receive a grant from the National Geographic Society to compare the floras and faunas of the exotic lovegrass plantations with those of nearby native grasslands. Were the plantations as rich in native plant or animal life as the adjacent uncolonized areas?

Between June 1984 and August 1985 we sampled vegetation, grasshoppers, birds, and rodents on thirty-six small plots dominated by African lovegrass, and compared these results with data from thirty-six plots in native grasslands (Table 5). In the end we found twenty-six native plants and animals to be more common on Research Ranch grassland mesas not planted with African lovegrasses, whereas no plants and only three native animals were more common in the exotic plantations. Without a doubt, the alien lovegrasses are bad news for most of the native grassland biodiversity in the Sonoita Valley, at least compared to relatively intact and undisturbed places.

The three animals more common in exotic lovegrass habitats are very interesting cases, because the other place where all three are most abundant on the sanctuary is in sacaton bottomlands. As we saw in Chapter 6, sacaton grass grows in tall, dense, highly uniform stands. Superficially at least, the dominant and lush growth of the African exotics,

FIGURE 33. A stand of Boer lovegrass on the Research Ranch, 1986. (Photo by the authors)

especially of Boer lovegrass, resembles that of sacaton. Bighead grasshoppers, Botteri's sparrows, and cotton rats apparently also recognized the similarity. It remains to be determined whether the resemblance between the two habitats is anything more than superficial. For example, we do not yet know how Botteri's sparrow nesting success compares between the two grasslands. Can this apparent habitat specialist successfully breed in an alien environment, or are the exotic grasslands an ecological trap?

In terms of native biodiversity, the only other potential benefit to exotic lovegrasses is that they may provide cover for some birds during drought years when stands of the native upland grasses are particularly sparse. For example, Caleb Gordon found that wintering grasshopper, Baird's, and vesper sparrows were more abundant on the sanctuary in exotic grasslands during a recent drought.

An old concept in ecology and conservation biology is that intact and undisturbed native communities should be able to resist invasions

TABLE 5. Native plants and animals more common in native grasslands on the Research Ranch, compared to those in stands of exotic African lovegrasses

Native Grasslands	*Exotic Grasslands*
Plants:	Plants:
blue grama	none
threeawn grass	
wolftail grass	
plains lovegrass	
burroweed	
groundsel	
copper leaf	
fleabane	
malvastrum	
caltrop	
Animals:	Animals:
striped grasshopper	largeheaded grasshopper
velvetstriped grasshopper	Botteri's sparrow
brownspotted grasshopper	cotton rat
crenulatewinged grasshopper	
Wyoming toothpick grasshopper	
whitewhiskered grasshopper	
barberpole grasshopper	
spurthroated grasshopper	
loggerhead shrike	
grasshopper sparrow	
Cassin's sparrow	
vesper sparrow	
eastern meadowlark	
western harvest mouse	
hispid pocket mouse	
northern pygmy mouse	

by exotic species. While this may be true, it has proven to be a very difficult hypothesis to test, since so little of the earth remains undisturbed. Certainly there are no grasslands in the American Southwest that qualify, given their nineteenth-century history of overgrazing. The question may have more hypothetical than practical significance in any event, since natural disturbances such as fires, floods, and landslides have always been, and will always be, a part of western American landscapes. The real issue is whether intact assemblages of native plants and animals will be *relatively* resistant to exotic invasions.

With regard to African lovegrasses on the Research Ranch, the jury is still out. Stands of Lehmann and Boer lovegrass are now easy to find along roadsides over much of the sanctuary, far beyond the bounds of their original plantations. It is clear that these species are more effective than most of the native grasses at colonizing bare soil. What is more alarming, however, is that the roadside populations have begun to spread out into adjacent native grasslands, particularly during drought years. Here again we see just how successful was the search for alien grasses from Old World deserts, perhaps with longer and stronger evolutionary histories of surviving droughts and fires than grasses native to the American Southwest.

The greatest challenge facing managers of the Research Ranch is finding ways to stop the spread of the alien lovegrasses. As we write this, there are no obvious or easy solutions. Both species are fire tolerant, and they may even increase in abundance following fire because of prolific seed production and the fact that the seedlings do particularly well in bare areas. In many grasslands, "weedy" species tend to have higher nitrogen requirements than native grasses, and so they can be inhibited by various methods that reduce soil fertility. We have no evidence that the African lovegrasses fall into this category. Like the native grasses, they are long-lived perennials whose growing seasons and ecological requirements appear generally similar to the natives. We are aware of no pathogens or herbivorous insects specific to the African lovegrasses that might be introduced as potential means of biological control. No herbicides exist that would kill the exotics while sparing the natives.

Yet there must be an answer. Because the alien grasses are genetically unique, by definition they have attributes making them distinct from all other species. The challenge is to discover which of those attributes might limit the exotics, under some as yet undiscovered set of environmental circumstances where native grasses, or at least most of them, might thrive. In all of conservation biology, there is no more important research need than finding ways to control the spread of exotic species. We are confident that solutions are out there.

Up to this point, we have written this chapter as if the evils of alien species are obvious and the need to control them self-evident. Yet those who would deflate the social and political influence of environmentalists and conservation biologists have seized upon this very subject as a major point of attack. What is the big deal about new species replacing old ones? After all, this sort of thing has been going on naturally for countless millennia as species evolve and spread, only to become extinct when something better comes along. *Homo sapiens* itself is a colonizing species, originally alien to most of the world, including all of the western hemisphere. Is it not fanciful and elitist to defend the insignificant and the meek against domination by naturally superior species? As long as the ecosystem continues to serve our needs, what difference does it make whether the Sonoita Plain is covered with grasses that were there at the time of Coronado or with some grasses imported in 1932 from southern Africa?

These are not trivial questions, and they deserve to be answered. Biologists, environmentalists, and philosophers offer a range of answers, sometimes passionately defending one or another as preeminent. Yet no single reason for conserving native biological diversity is exclusive or incompatible with any other, and so they all can and should be considered. Full exposition on this subject would require a much longer book and more wisdom than we alone can offer. Instead, we provide a simple (and doubtless incomplete) list of answers to the following question: *Why would it be unfortunate if native grasslands in the Sonoita Valley were to be displaced by lovegrasses native to southern Africa?*

1. We would lose things we need right now, including plants we can eat and wildlife we can hunt. Recall that exotic plantations are substantially devoid of native animals as well as vegetation.

2. We would lose things we will need in the future. Native plants and animals in the Sonoita Valley represent unique genetic resources of immeasurable potential value, for food, fiber, and medicine.
3. Species-poor ecosystems are more likely to fail as providers of essential ecosystem services, such as soil stabilization, watershed protection, and moderation of atmospheric conditions (e.g., carbon storage as a deterrent to possible global warming). If a key grass or herbivore or predator declines, for whatever reason, there will be few if any equivalent species around to fill in and provide the same ecosystem services—a situation known as limited species functional redundancy.
4. We have a moral obligation to protect the full variety of living things on Earth. We were not responsible for their creation, and we have no right to cause their extinction, however often that might occur as part of the natural order of things. African lovegrasses are in no danger of extinction, but the same cannot be said for many species of plants and animals found only in arid grasslands of the American Southwest—especially those threatened by the spread of exotics.
5. The Sonoita Valley would be a much less interesting place to live, absent the rich variety of plants and animals found there today. Furthermore, it would no longer be distinctive from all the other places where the African lovegrasses have already spread.

While these answers are disparate, in the end they all converge on a single theme we have raised elsewhere in this book. One of the most striking impacts of humans is to homogenize the Earth—to reduce its variety and diversity. Lehmann lovegrass is no less aesthetic or environmentally valuable than any other grass, and there would be little consequence or concern if it simply integrated in as part of the variety of plants and animals already living in the Sonoita Valley. But that is not what happens.

Most field biologists we know mourn the loss of diversity of places, because it was the variety of places and their living things that attracted them to biology in the first place. Yet the desire for variety, and heterogeneity, and distinctiveness of place—especially places that are natural—must extend to humans in general. How else can we explain the overwhelming popularity of our national parks and other open spaces?

It is difficult to predict when or even if, by virtue of our numbers and our greed, we might cause irreparable damage to Earth's life-support system. Certainly it is wise to proceed with caution in the meantime. We suspect, however, that the more imminent danger is to our spiritual well-being, as we homogenize the world.

Fourteen

THE WORLD IS FULL OF LIFE

For a variety of personal and professional reasons, we took a break from the Research Ranch at the end of the 1991 field season. We thought at the time that it might have been for keeps, but the lure of the place proved too strong. We have come back, though in a different capacity. The simplest way to describe it is to say that we promoted ourselves from parents to grandparents. We still care deeply about the place, and we have some personal research projects going. But now it is somebody else's problem to balance the budget, to fund the research, and to fill the dormitory with good field scientists. This might sound like laziness or arrogance, but it is neither. It is recognition that no individual, or pair of individuals, has all the good ideas or knows all the right ways. A great many worthy institutions have suffered because somebody could not or would not see when it was time to promote themselves into the role of grandparents.

The National Audubon Society has recruited some world-class people to care for the Research Ranch (Fig. 34), to learn more about the ecological forces at work there, and to share that information with others. Our first return visit was in the summer of 1995. The place was in wonderful shape, but there were more lights on the night sky and more

FIGURE 34. Bill Branan, manager of the Research Ranch, 1998. (Photo by Erika Geiger)

houses scattered across the Sonoita Plain. Even Elgin was showing signs of resurgence. The Ranch seemed less isolated, the views more threatened. As Bill Brophy had predicted, there was more discussion about comparing the sanctuary with housing developments than with ranches. It all made us uneasy. Later, flying out of Tucson, that unease spawned some reflective conversation and the following thoughts, with which we end this book.

One consequence of technology and urbanization—ultimately, the one for which we shall pay the highest price—is that more and more Americans are becoming disconnected from what is left of the natural world. They will miss the beauty, and that is bad enough, but they will also avoid contact with their own roots, and with some important lessons in ecological reality.

A conservative New York radio celebrity wrote a book a few years ago entitled *The Way Things Ought To Be.* We are neither wise nor foolish enough to write a whole book entitled *The Way Things Actually Are*, but we do know that one of its essential chapters should be called "The World Is Full of Life."

Then again, perhaps a better title would be "The World Is *Already* Full of Life." It is an irrefutable fact of biology that all living things—including humans—have a genetically predetermined capacity to overwhelm the universe with their own numbers, and to do it quickly. That is *why* the world is full of life. All opportunities are quickly filled because there is always some organism out there whose offspring need the space. Lots of landscapes in the American West may look empty, but that is an illusion. Even here all the good spots are being used by something—*right now.*

All living things survive only at the expense of some other living things, either because those other living things are food or, equally often, because they are the competition. These facts are neither new nor obscure, but they are easily ignored or forgotten unless you live in the natural world and are watchful. These facts also undermine some of our most cherished urban myths, the central and most dangerous one being that all the human things we care most about—our families and our institutions—can keep on getting bigger, forever.

A debate is currently raging about the seriousness of global human population growth. No rational person could deny that there must be some upper limit to our numbers, given that the earth is finite. Most people also would agree that there is a negative relationship between the numbers of us trying to occupy the globe and the quality of our individual lives. The debate centers on what, if anything, to do about it.

One view holds that strident calls for global population control are simplistic and thinly disguised attempts to perpetuate economic, sexist, and racist policies invented by white males to retain their world dominance. According to this view, once racial, gender, and cultural equalities become real, the population problem somehow will take care of itself.

We are skeptical. Perhaps it is because we live in a place in Colorado where most of the people are wealthy (obscenely wealthy by world standards), where racial and gender relationships are reasonably good, and where the mostly well-to-do and mostly white human population is growing absolutely out of control. It matters little to us that this growth is due largely to immigration from other parts of the United States rather than being the result of local reproduction. All those people are coming from somewhere, and *those* places would be a mess if there weren't so many people leaving.

Certainly there can be no environmental justice without social justice. Humans who are forced to live in an unjust world have neither the time nor the inspiration to contemplate how their descendants might someday live in a manner that sustains both themselves and the natural world over the long term. Solving environmental problems needs to involve everybody, in arenas where all opinions and needs have equal weight. But we are not convinced that social and economic reform will suffice. Has there ever been a society in the history of the world that, when presented with a technology that enhances its numbers and allows suppression of the natural world, has chosen not to use it? Norman Maclean's observation that humankind is fundamentally a mess would appear to transcend all boundaries of race, gender, and culture.

We suspect that living in true harmony with the natural world, in a manner sustainable over the long run, is something no modern human society has yet learned how to do. The survival of the natural world, however, and likely our survival as a species, depends on our learning to do this. It will be a unique experience in human history. Most thoughtful ecologists and conservation biologists are not sure it is possible. That is why so many of us are grumps, why some few of us become anarchists, and why some of us just go fishing and skiing.

Another urban myth holds that humans can live in the world and do no harm to it as long as they are sufficiently virtuous. For example, some individuals take pride in the fact that they live without killing other animals. Now, if we eat hamburgers made from steers that grazed on the grasslands of the Sonoita Valley, we certainly have lived at the expense of some grassland plants and animals, but probably not at the expense of all of them. On the other hand, if we convert the Arizona high plains into soybean fields and vineyards, the grasslands are lost. All the native plants and animals that used to be there are dead and gone forever. They can't just move to some other place, because that place is already taken.

We can live only at the expense of other lives, but this selfishness is not a human attribute, it is an inescapable characteristic of all living things. What is uniquely human is not our striving to numerically dominate the world, but our historic success. We should not despair or stop living, but neither should we live without regard for the consequences of our collective abundance or our individual behavior. There is no ethical paradox here. Life is too good to pass up; but ever-increasing human life, not shared with the natural world, becomes ecologically doomed at some point, and spiritually bankrupt long before then.

For billions of years the rules of the game have been that whoever has the most kids wins. We are an evolutionary consequence of that game, and now we are so good at it that we are overwhelming the world, simply crushing it with our collective numbers and greed. Coming to grips with the reality of population growth is so horrifying that we will go to almost any lengths to deny it. We point to the fact that most of the world is still "unoccupied." We pontificate that limiting our own abundance is somehow immoral, instead of simply being contrary to our inherent reproductive drives. We point to our undeniable skill at finding new technologies and pronounce those urging restraint as doomsayers. We forget that the world already is full of life, and we kill the messenger who reminds us.

It may be bitter medicine, but if we are to survive ecologically and spiritually, we must now change the rules of the game. We must come to our reproductive senses before it is too late. Meanwhile, those of us who cannot live without a view have a choice. We can run away, at once looking over our shoulders and looking ahead, hoping to find one more place that has not yet been obliterated by the tide. Or we can draw a line across the plains and the savannas we love so much, and take a stand.

FIGURE 35. The windmill on Bald Hill, 1982. (Photo by the authors)

Appendix

SCIENTIFIC AND ENGLISH NAMES OF PLANTS AND ANIMALS

GRASSES[1]

**Agropyron desertorum*	crested wheatgrass
Aristida[2] spp.	threeawn grasses
Bothriochloa barbinodis	cane beardgrass
Bouteloua chondrosioides	sprucetop grama
Bouteloua curtipendula	sideoats grama
Bouteloua eriopoda	black grama
Bouteloua gracilis	blue grama
Bouteloua hirsuta	hairy grama
**Bromus inermis*	smooth brome
Digitaria californica	Arizona cottontop
Elyonurus barbiculmis	woolly bunchgrass

Note: Species marked with an asterisk are referred to in the text but have not been recorded at the Research Ranch.

1. Where available, plant names are taken from McClaran and Van Devender 1995; Turner, Bowers, and Burgess 1995; and Bowers and McLaughlin 1996; otherwise from Kearney and Peebles 1960.

2. Present taxonomy of the genus *Aristida* is complex and uncertain. Species identified on the sanctuary include *A. purpurea, A. longiseta,* and *A. divaricata.* See J. H. Bock and C. E. Bock 1986.

Eragrostis curvula[3]	Boer lovegrass
Eragrostis intermedia	plains lovegrass
Eragrostis lehmanniana	Lehmann lovegrass
Heteropogon contortus	tanglehead
Hilaria belangeri	curly mesquite
Lycurus phleoides	wolftail
Panicum obtusum	vine mesquite
Schizachyrium cirratus	Texas beardgrass
Sporobolus wrightii	big sacaton
Trachypogon secundus	crinkle-awn grass

HERBS

Acalypha lindheimeri	copper leaf
Erigeron divergens	fleabane
Kallstroemia parviflora	caltrop
Malvastrum bicuspidatum	malvastrum
Zinnia acerosa	white zinnia

TREES, SHRUBS, AND SUCCULENTS

Acacia spp.	acacia
Acer grandidentatum	big-tooth maple
Agave palmeri	agave (lechuguilla)
Baccharis pteronioides	yerba de pasmo
Chilopsis linearis	desert willow
Fouquieria splendens	ocotillo
Fraxinus pennsylvanica (ssp. *velutina*)	velvet ash
Isocoma tenuisecta	burroweed
Juglans major	Arizona walnut
Juniperus deppeana	alligator juniper
Mimosa aculeaticarpa	catclaw mimosa
Mimosa dysocarpa	velvet-pod mimosa
Platanus wrightii	Wright's sycamore
Populus fremontii	Fremont cottonwood
Prosopis velutina	mesquite
**Pseudotsuga menziesii*	Douglas fir

3. Boer lovegrass is considered either as *Eragrostis chloromelas* (e.g. Gould 1951) or, more recently, as a variety of weeping lovegrass (*E. curvula* var. *conferta;* see e.g. McClaran and Van Devender 1995).

Quercus arizonica	Arizona white oak
Quercus emoryi	Emory oak
Salix spp.	willow
Senecio douglasii	groundsel
Tamarix ramosissima	tamarisk

GRASSHOPPERS[4]

Ageneotettix deorum	whitewhiskered grasshopper
Amphitornus coloradus	striped grasshopper
Cordillacris crenulata	crenulatewinged grasshopper
Dactylotum variegatum	barberpole grasshopper
Eritettix simplex	velvetstriped grasshopper
Melanoplus desultorius	spurthroated grasshopper
Parapomala wyomingensis	Wyoming toothpick grasshopper
Phoetaliotes nebrascensis	largeheaded grasshopper
Psoloessa texana	brownspotted grasshopper

AMPHIBIANS AND REPTILES[5]

Crotalus atrox	western diamondback rattlesnake
Crotalus molossus	black-tailed rattlesnake
Crotalus scutulatus	Mojave rattlesnake
Gopherus flavomarginatus	Bolson tortoise
Rana chiricahuensis	Chiricahua leopard frog
Sceloporus slevini	bunchgrass lizard

BIRDS[6]

Accipiter cooperii	Cooper's hawk
Aimophila botterii	Botteri's sparrow
Aimophila cassinii	Cassin's sparrow
Aimophila ruficeps	rufous-crowned sparrow
Ammodramus bairdii	Baird's sparrow
Ammodramus savannarum	grasshopper sparrow

4. Nomenclature from Pfadt 1994.

5. Nomenclature from J. T. Collins 1997, and for the bunchgrass lizard, from Smith et al. 1996.

6. Nomenclature from American Ornithologists' Union 1983 and supplements published in the *Auk*.

Aphelocoma ultramarina	Mexican jay
Bombycilla cedrorum	cedar waxwing
Bubo virginianus	great horned owl
Buteo jamaicensis	red-tailed hawk
Calcarius ornatus	chestnut-collared longspur
**Cardinalis cardinalis*	northern cardinal
Carpodacus mexicanus	house finch
Circus cyaneus	northern harrier
Colaptes auratus	northern flicker
Contopus sordidulus	western wood-pewee
Corvus cryptoleucus	Chihuahuan raven
**Cyanocitta cristata*	blue jay
Cyrtonyx montezumae	Montezuma quail
Dendroica graciae	Grace's warbler
Dendroica petechia	yellow warbler
Eremophila alpestris	horned lark
Falco mexicanus	prairie falcon
Falco sparverius	American kestrel
Geothlypis trichas	common yellowthroat
Guiraca caerulea	blue grosbeak
Icterus bullockii	Bullock's oriole
Lanius ludovicianus	loggerhead shrike
**Melanerpes erythrocephalus*	red-headed woodpecker
Melanerpes formicivorus	acorn woodpecker
Myadestes townsendi	Townsend's solitaire
Myiarchus cinerascens	ash-throated flycatcher
Myiarchus tuberculifer	dusky-capped flycatcher
Parus wollweberi	bridled titmouse
Passerculus sandwichensis	Savannah sparrow
Pheucticus melanocephalus	black-headed grosbeak
Pipilo fuscus	canyon towhee
Piranga flava	hepatic tanager
Pooecetes gramineus	vesper sparrow
Salpinctes obsoletus	rock wren
Sialia currucoides	mountain bluebird
Sialia mexicana	western bluebird
Sialia sialis	eastern bluebird
Sitta carolinensis	white-breasted nuthatch
Spizella passerina	chipping sparrow
Sturnella magna	eastern meadowlark
Thyromanes bewickii	Bewick's wren
Turdus migratorius	American robin
Tyrannus vociferans	Cassin's kingbird

Vermivora luciae	Lucy's warbler
Zenaida macroura	mourning dove
Zonotrichia leucophrys	white-crowned sparrow

MAMMALS[7]

Antilocapra americana	pronghorn
Baiomys taylori	northern pygmy mouse
**Bison bison*	bison
Canis latrans	coyote
Canis lupus baileyi[8]	Mexican wolf
Chaetodipus hispidus	hispid pocket mouse
Cynomys ludovicianus[8]	black-tailed prairie dog
Dipodomys merriami	Merriam's kangaroo rat
Erethizon dorsatum	porcupine
Lepus californicus	black-tailed jackrabbit
Lynx rufus	bobcat
Neotoma spp.	woodrat
Odocoileus hemionus	mule deer
Odocoileus virginianus	white-tailed deer
Peromyscus maniculatus	deer mouse
Reithrodontomys megalotis	western harvest mouse
Sciurus arizonensis	Arizona gray squirrel
Sigmodon arizonae[9]	Arizona cotton rat
Sigmodon fulviventer	tawny-bellied cotton rat
Spermophilus variegatus	rock squirrel
Sylvilagus audubonii	desert cottontail
Taxidea taxus	American badger
Tayassu tajacu	collared peccary (javelina)

7. Nomenclature from J. K. Jones et al. 1992.

8. Not on the Research Ranch, but doubtless present in former times.

9. We identified these rodents as *Sigmodon hispidus* in our earlier work at the sanctuary (e.g., C. E. Bock and J. H. Bock 1978), but recent studies indicate they are a distinct species, *S. arizonae* (see Hoffmeister 1986). This is the common species in sacaton bottomlands, whereas *S. fulviventer* is more abundant in uplands.

NOTES

The literature related to the topics discussed in this book is far too extensive for a complete review. Our limited goals for each chapter therefore are three: to provide the reader with access to particularly important and synthetic works on each subject; to list all of the works conducted on the Research Ranch focused on the topic; and to give some examples of recent case studies conducted elsewhere.

CHAPTER 1. THE GRASSLANDS OF CORONADO

Conrad Bahre (1977) wrote a land-use history of the Research Ranch, beginning with the time of prehistoric peoples on the Sonoita Plain. This work was followed by more general accounts of human impacts on grasslands and other environments of the Arizona borderlands, with particular emphasis on livestock grazing, fire suppression, wood cutting, and introductions of exotic species (Bahre 1991, 1995; Bahre and Shelton 1993, 1996). Early vegetation studies at the Research Ranch were conducted by Bonham (1972) and S. White (1974).

Other important studies of natural history, vegetation change, and human impacts and history in southeastern Arizona are Hastings and Turner 1965; Gehlbach 1981; Humphrey 1987; and, more generally, Limerick 1987, 1989. Axelrod (1985); Betancourt, Van Devender, and Martin (1990); and Van Devender (1995) analyzed the paleo-ecological history of North American grasslands generally. Van Devender concluded that contemporary grasslands of the American Southwest have been in place only for about the last nine thousand years.

Southwestern grasslands are bounded at their lower elevational limits by Sonoran and Chihuahuan desert scrub, and at their upper elevational limits by woodlands and forests of montane "sky islands" at the northern end of the Sierra Madre Occidental (D.E. Brown 1982; J.P. Wilson 1995). Grasslands of the Sonoita Valley have been variously described as desert grassland, semidesert grassland, high desert sod-grass, or plains grassland (D.E. Brown 1982, Schmutz et al. 1991, McClaran 1995). As on the Research Ranch, these grasslands intergrade at higher elevations of surrounding hills and mountains into oak-dominated Madrean evergreen woodlands (D.E. Brown 1982). The whole area is classified as part of a "Madrean Floristic Province" (McLaughlin 1992) because of its strong overall affinities with the upland vegetation of northwestern Mexico.

CHAPTER 2. ON BEING THE CONTROL

Most southwestern grasslands support complex mixtures of scattered trees, shrubs, succulents, herbs, and perennial grasses that vary in relation to elevation, temperature, precipitation, soils, and disturbance history (Burgess 1995), and endemic animals appear to respond positively to this heterogeneity (Whitford 1997). Turner et al. (1980) assembled information about various long-term livestock exclosures in Arizona where these patterns might be studied without the confounding influence of contemporary grazing.

Paleoecological evidence reveals that there have been episodic and reversible changes in the abundance of grasses versus woody plants over the past nine thousand years in many lowland southwestern sites, probably in response to changes in climate (Van Devender 1995). By contrast, historic livestock grazing and accompanying changes in land use have caused essentially permanent shifts from grassland to shrubland, by selective removal of grasses, reductions in fire frequency, alterations in soil structure and patterns of soil nutrient distribution, dispersal and planting of seeds of woody plants by livestock, and prairie dog control (Buffington and Herbel 1965; Tiedemann and Klemmedson 1973; Dobyns 1981; Walker et al. 1981; Chew 1982; Schlesinger et al. 1990; Bahre and Shelton 1993, 1996; Archer 1994; National Research Council 1994; Weltzin, Archer, and Heitschmidt 1997; Whitford 1997; Kramp, Ansley, and Tunnell 1998).

It is important to remember that both grasses and woody plants of the southwestern desert grasslands are native to the region and have evolutionary associations with it. They are likely to be better adapted to periodic droughts and fires, and to grazing and browsing by native herbivores over the past few millennia, than to the effects of introduced livestock that have been around in large numbers only for the past 150 years.

At higher and somewhat wetter sites, such as the Sonoita Valley, grasslands survived the intense grazing and droughts of the 1890s but probably emerged with higher densities of trees and shrubs such as mesquite, catclaw mimosa, yerba de pasmo, and burroweed. Burroweed in particular seems to respond positively to the independent and interactive effects of fire suppression and episodic increases in winter precipitation (S.C. Martin 1975; C.E. Bock and J.H. Bock 1997).

Roundy and Biedenbender (1995) provide a comprehensive review of historic and current attempts to revegetate degraded southwestern desert grasslands. To date, reseeding with native perennial grasses usually has failed, unless it has been coupled with more expensive and intensive management strategies designed to increase seedling survival, such as plowing and contouring to increase water infiltration, and mechanical or chemical brush control. Seeding exotic lovegrasses has been much more successful, but apparently at the expense of the remaining native flora and fauna (see Chapter 13).

A comparatively recent approach to improving the condition of southwestern rangelands, as well as livestock production, is "Holistic Resource Management" (HRM; Savory and Butterfield, 1999). The centerpiece of HRM is high-density, short-duration rotational grazing. This method of cattle management requires livestock to be fenced into small paddocks, which encourages them to consume all vegetation more uniformly, including those plants they might ignore without such confinement. The technique also is purported to increase soil fertility via livestock urine and feces and to enhance soil water infiltration by means of the action of their hooves. The effects of HRM on vegetation, soils, and livestock production have been tested under a variety of circumstances, with decidedly mixed, mostly negative, findings (e.g., Weltz and Wood 1986; Heitschmidt, Dowhower, and Walker 1987; M. R. White et al. 1991; Taylor, Brooks, and Garza 1993). Rangelands managed using the principles of Holistic Resource Management rarely have been compared to large, long-term livestock exclosures.

To reiterate: the value of the Research Ranch in the context of restoration and revegetation is not that it is somehow "better" than other desert grasslands. Rather, its value is that it can serve as a yardstick against which the consequences of intensive management of similar areas can be measured and evaluated. See Arcese and Sinclair 1997 for a recent discussion of the value of such protected areas as ecological baselines.

CHAPTER 3. WAITING FOR RAIN

Neilson (1986, 1987) analyzed the dynamics of the summer monsoon rainy season in the American Southwest, along with patterns of winter rain, and their comparative effects on desert grass and shrubland vegetation. Bahre and Shelton (1993, 1996) reviewed the combined impacts of grazing and droughts on desert grasslands at the end of the nineteenth century. Elevated atmospheric carbon dioxide levels and nitrogen deposition related to industrialization almost certainly will affect world vegetation (Neilson and Marks 1995; Korner and Bazzaz 1996). Increased carbon dioxide may favor expansion of woody shrubs and trees into southwestern desert grasslands, as well as causing other vegetation changes (e.g., Guo and Brown 1997; Polley 1997; Polley et al. 1997; Golluscio, Sala, and Lauenroth 1998; Jackson et al. 1998). However, it will be very difficult to separate climate and carbon dioxide effects from the more immediate influences of livestock grazing and fire suppression (Archer, Schimel, and Holland 1995).

Joseph Grinnell (1922) was among the first to recognize the ecological and evolutionary importance of dispersal of animals into previously or temporarily unoccupied habitats. J. H. Brown (1995) recently elaborated on this theme in a general treatise on the distribution and abundance of species.

We examined habitat relations of the javelina (collared peccary) on the Research Ranch in 1976–1977 (C. E. Bock and J. H. Bock 1979). For more general works on the distribution, ecology, and social behavior of this species, see Bissonette 1982; Byers and Bekoff 1981; Sowls 1984; and Hellgren et al. 1995.

Dunning and Brown (1982) documented the relationship between the preceding summer's rainfall and birds wintering in the grasslands of southeastern Arizona. Ron Pulliam and his students examined the importance of summer rain and resulting grassland seed production to winter sparrow abundance and survival on the Research Ranch sanctuary (e.g. Pulliam and Brand 1975; Pulliam and Dunning 1987), while C. E. Bock and J. H. Bock (1999) compared winter bird abundances on and adjacent to the sanctuary during and after a drought.

CHAPTER 4. FENCELINES

Bahre (1977, 1991, 1995, 1998) reviewed the historic impacts of livestock grazing and wood cutting in the Sonoita Valley and elsewhere in southeastern Arizona. There is much opinion on the contemporary impacts of livestock grazing on grasslands of the American West, including the Southwest. According to some (e.g. Savory and Butterfield, 1999), grazing by large ungulates is essential to the maintenance of healthy rangelands. Others point to its continuing negative impacts on vegetation, wildlife, soils, and watersheds (e.g. Ferguson and Ferguson 1983; Jacobs 1992). The subject remains highly controversial, probably both because of a lack of sufficient study and owing to its cultural and political aspects (e.g., Brussard, Murphy, and Tracy 1994; Noss 1994; Wuerthner 1994; J. H. Brown and McDonald 1995; Fleischner 1996).

There is a substantial literature on the effects of livestock grazing on vegetation and wildlife in grasslands of the American West. A comprehensive review is beyond the scope of this book, but recent summaries of the ecological effects of livestock grazing on western rangelands include Fleischner 1994; Vavra, Laycock, and Pieper 1994; Saab et al. 1995; Krausman 1996; and Milchunas, Lauenroth, and Burke 1998. Domestic livestock can impact any grassland if stocked in sufficient numbers; however, they have had much greater negative effects in ecosystems that have not had a recent evolutionary association with large herds of bison, such as grasslands of the Great Basin, California, and the desert Southwest (McDonald 1981; Mack and Thompson 1982; Painter and Belsky 1993; Holechek et al. 1994; Painter 1995; Truett 1996; Tracy, Golden, and Crist 1998). By contrast, grasslands of the western Great Plains, where bison were abundant, have proven more tolerant of the impacts of livestock grazing, probably because the dominant grasses have evolved structural and functional attributes that make them relatively resistant to herbivory (Milchunas, Sala, and Lauenroth 1988; Milchunas, Lauenroth, and Burke 1998).

Cross-fence and post-grazing studies at the Research Ranch have included comparisons of perennial grasses (C. E. Bock et al. 1984; Brady et al. 1989; C. E. Bock and J. H. Bock 1993), shrubs (Kenney, Bock, and Bock 1986; C. E. Bock and J. H. Bock 1997), grasshoppers (Jepson-Innes and Bock 1989), lizards (C. E. Bock, Smith, and Bock 1990), and birds and rodents (C. E. Bock et al. 1984; C. E. Bock and Webb 1984; C. E. Bock and J. H. Bock 1988, 1999; see also M. R. Stromberg 1990). J. H. Bock and C. E. Bock (1986) described habitat relationships of major grasses on the sanctuary.

The comment about Hayduke refers to a character in Edward Abbey's novel *The Monkey Wrench Gang* (1975).

CHAPTER 5. HERPS

The concept of indicator species is an old one (Merriam 1898; Hall and Grinnell 1919; Clements 1920), and there have been recent criticisms of its value and application (Landres, Verner, and Thomas 1988; Murtaugh 1996; Dufrene and Legendre 1997; Niemi et al. 1997). Nevertheless, reptiles and amphibians, along with certain plants, may be a good candidate group because their life histories are well known, they are relatively immobile (compared, say, to birds), and they frequently have rather specific habitat requirements.

Stebbins's field guide (1985) provides excellent basic information on the habitats and distribution of reptiles and amphibians of the western United States.

There have been few studies of herps in relation to livestock grazing in the Southwest (but see K. B. Jones 1981, 1988). C. E. Bock, Smith, and Bock (1990) quantified the great abundance of bunchgrass lizards on the Research Ranch in 1989, compared to adjacent grazed lands. However, by 1997 this species had virtually disappeared from the sanctuary and surrounding areas, perhaps owing to the severe drought of 1995–1996 (Smith et al. 1998). Recently, Ballinger and Congdon (1996) found a significant decline in bunchgrass lizards in the Chiricahua Mountains of extreme southeastern Arizona, apparently as a result of livestock grazing–induced reductions in grass cover.

Guillette et al. (1980) reviewed the correlation between high-elevation and cold-climate distribution and the trend toward viviparity (live-bearing) in the genus *Sceloporus,* to which the oviparous (egg-laying) bunchgrass lizard may be an exception (see also Mathies and Andrews 1995). For a recent perspective on evolution in this group of lizards, including issues of the evolution of viviparity, see Creer et al. (1997).

Mendelson and Jennings (1992) described the decline in Mojave rattlesnakes and the increase in western diamondback rattlesnakes associated with desertification of former grasslands along the southernmost border between Arizona and New Mexico. Most ecologists now agree that complementary distribution patterns and abundances of species (i.e., one species replacing another in space or time), such as those of the three rattlesnakes on the Research Ranch, provide only circumstantial evidence for interspecific competition, unless accompanied by controlled field experiments (e.g. Schoener 1982; Connor and Simberloff 1986; Wiens 1989).

Recent declines in amphibian populations, and their possible causes, are reviewed by Blaustein and Wake (1995); Pechmann and Wake (1997); and Wake (1998).

CHAPTER 6. ISLANDS OF FIRE

There are four comprehensive reviews about the effects of fire in natural ecosystems in North America, by Kozlowski and Ahlgren (1974); Wright and Bailey (1982); Pyne, Andrews, and Laven (1996); and DeBano, Neary, and Ffolliott (1998). R. C. Anderson (1982) proposed a model describing the interactive importance of drought, fire, and grazing in determining the distribution and function of grasslands. Bragg (1995) reviewed the importance of fire across the North American Great Plains as a whole, while Bragg and Hulbert (1976); Gibson and Hulbert (1987); Vinton et al. (1993); and Knapp et al. (1998) described the specific effects of fire and other ecological forces at work on the Konza tallgrass prairie.

Aldo Leopold (1924) and subsequently Robert Humphrey (1958) were among the first to consider the importance of fire in desert grasslands, while more recent reviews on this topic are found in Dobyns 1981; Bahre 1985, 1991; Cox, Ibarra F., and Martin 1990; Krammes 1990; and McPherson 1995. Archer (1994) and Archer, Schimel, and Holland (1995) reviewed the history and possible causes of woody plant encroachment into southwestern grasslands. Weltzin, Archer, and Heitschmidt (1997) described the role that black-tailed prairie dogs, and other animals associated with their colonies, may once have played in preventing woody plants from invading and increasing in North American grasslands. Kramp, Ansley, and Tunnell (1998) analyzed the importance of livestock and wildlife in dispersing seeds of mesquite and other woody plants into arid southwestern grasslands. Maurer (1985), Whitford (1997), and Lloyd et al. (1998) noted the importance of woody plants to a variety of birds and small mammals in desert grassland ecosystems, which would argue for a habitat mosaic including some areas protected from fire.

Results of fire studies at the Research Ranch were reported in the following publications: J. H. Bock, Bock, and McKnight 1976; C. E. Bock and J. H. Bock 1978, 1979, 1991, 1992, 1997; Thomas and Goodson 1992; C. E. Bock et al. 1995; and J. H. Bock and C. E. Bock 1992a, 1992b.

CHAPTER 7. OAKS, ACORNS, AND RUGGED GROUPS

E. O. Wilson (1992) and Kellert and Wilson (1993) described the biophilia hypothesis and its relationships to species diversity and conservation, while McPherson (1997) synthesized information about the ecology and management of North American savannas.

Sanchini (1981) compared the demographics of Emory oak and Arizona white oak on the Research Ranch. For information about acorn crops, their tannins, and the role of animals in acorn dispersal and planting, see Koenig 1991; Steele et al. 1993; Koenig et al. 1994; Fleck and Tomback 1996; Fleck and Woolfenden 1997; J. A. Hubbard and McPherson 1997; W. C. Johnson et al. 1997; and Koenig and Faeth 1998. Koenig and Knops (1998) documented the geographic scale of masting among certain trees species and the potential ecological and evolutionary consequences of that scale.

Ortega (1987b, 1990a, 1990b, 1991) studied rock squirrel ecology and behavior at the Research Ranch. Stacey and Koenig (1990) edited a volume on cooperative breeding in birds, including the Mexican jay and acorn woodpecker. Studies of the Mexican jay at the Southwestern Research Station are summarized in J. L. Brown and Brown 1990; J. L. Brown 1994; and J. L. Brown et al. 1997. Leach (1925) was one of the first to describe communal living in the acorn woodpecker. Our work on this species at the Research Ranch was presented in Stacey and Bock 1978. For general information about acorn woodpecker behavioral ecology, including work at the Hastings Reservation in California, see Stacey and Koenig 1984; Koenig and Mumme 1987; and Koenig, Haydock, and Stanback 1998.

C. E. Bock and J. H. Bock (1979) examined habitat relationships of collared peccaries (javelina) on the Research Ranch, while Byers and Bekoff (1981); Bissonette (1982); and Hellgren et al. (1995) studied various aspects of the social behavior and ecology of peccaries in central Arizona and southern Texas.

CHAPTER 8. LITTLE BROWN BIRDS

The Birds of Arizona by Phillips, Marshall, and Monson (1964) provides good general information about sparrows and other species associated with southwestern grasslands. D. E. Brown and Davis (1998) reviewed historical changes in birds and mammals of the American Southwest. See W. A. Davis and Russell 1995 for an annotated checklist of birds in southeastern Arizona. Goriup (1988) edited a book on the ecology and conservation of grassland birds of the world. The Breeding Bird Survey (BBS) of the Biological Resources Division of the U.S. Geological Survey provides long-term data on the distribution and abundance of summer birds in North America, while analogous winter information is compiled in the Audubon Society's annual Christmas Bird Count (Root 1988; Price, Droege, and Price 1995). BBS data suggest that a variety of grassland birds are undergoing substantial declines, at least in certain areas (Robbins, Sauer, and Peterjohn 1993; Knopf 1994). Birds may be among the most susceptible animals to the ecological effects of livestock grazing (Milchunas, Lauenroth, and Burke 1998).

Vickery (1996) summarized the biology of grasshopper sparrows. Our work on this species at the Research Ranch was published in C. E. Bock et al. 1984 and C. E. Bock and Webb 1984. For a recent study of the biology of the Baird's sparrow, see S. K.

Davis and Sealy 1998. Webb and Bock (1996) compiled information about the Botteri's sparrow, with additional information about its particular association with sacaton reported in Webb and Bock 1990. J. P. Hubbard (1977) and Dunning et al. (1999) reviewed the biology of the Cassin's sparrow.

Dunning and Brown (1982) analyzed the relationship between rainfall, seed production, and the abundance of sparrows wintering in the Southwest. Ron Pulliam and his students investigated winter distribution and abundance of sparrows at the Research Ranch, in relation to grassland seed production and preferred habitats (e.g., Pulliam 1975, 1983, 1985; Pulliam and Brand 1975; Pulliam and Mills 1977; Pulliam and Dunning 1987; see also C. E. Bock and J. H. Bock 1999). Lima and Valone (1991) studied winter distribution of sparrows on the sanctuary in relation to escape cover and predator avoidance strategies. In these and other cases worldwide there is solid evidence that densities of grassland sparrows and finches are limited by seed abundance but that the threat of predation also is important to sparrow habitat selection (Schluter and Repasky 1991).

Karen Jepson-Innes first quantified the abundance of grasshoppers on and adjacent to the Research Ranch (Jepson-Innes and Bock 1989) and made initial observations on the importance of these insects in the diets of breeding Cassin's sparrows. C. E. Bock, Bock, and Grant (1992) documented the increase in grasshopper densities on the sanctuary following experimental exclusion of songbirds. Other experimental studies quantifying the impacts of avian predators on insects include Holmes, Schultz, and Nothnagle 1979; Wiens et al. 1991; Joern 1992; Belovsky and Slade 1993; and Floyd 1996.

CHAPTER 9. FRAGMENTS

Several major symposia have been devoted to the ecology and management of riparian ecosystems in the western United States (e.g., R. R. Johnson and Jones 1977; R. R. Johnson et al. 1985). There is a major emphasis on conservation of these ecosystems because of the productivity and diversity of plant and animal populations that they support and because of the variety of threats to their ecological integrity (Knopf et al. 1988; Krueper 1994; Ohmart 1994; Dobkin, Rich, and Pyle 1998).

Barstad (1981) and Cross (1991) studied the distribution of vegetation in riparian habitats on the Research Ranch and elsewhere in the Sonoita Valley. We examined patterns of distribution and reproduction among populations of Wright's sycamore in the Sonoita Valley, in the Huachuca Mountains, and elsewhere in southeastern Arizona (J. H. Bock and C. E. Bock 1985, 1989). For a recent example of the consequences of groundwater depletion on Arizona riparian ecosystems, see J. C. Stromberg, Tiller, and Richter 1996.

Western riparian habitats have been recognized in particular for their abundance and variety of bird life (Terbourgh 1989; Saab et al. 1995). Riparian corridors in southwestern lowlands have been especially well studied, and are known to support the highest densities of breeding birds of any habitat in North America (Carothers, Johnson, and Aitchison 1974; B. W. Anderson and Ohmart 1977; Stamp 1978; Rice, Anderson, and Ohmart 1984; Szaro and Jakle 1985). They also are important for birds during migration, where small oases of riparian trees may be just as important as larger stands (Skagen et al. 1998). Results of our own work on riparian birds in the Sonoita Valley and vicinity were reported in C. E. Bock and J. H. Bock 1984 and T. R. Strong and Bock 1990.

The principles of landscape ecology are presented in Forman and Gordon 1986. Applications of landscape ecology and sink-source population dynamics to conservation biology, and to population and community ecology in general, are reviewed in Soulé 1986; Pulliam 1988; Dunning et al. 1995; Freemark et al. 1995; Kareiva and Wennergren 1995; Brawn and Robinson 1996; McCullough 1996; Hanski and Gilpin 1997; and Diffendorfer 1998.

Andrén (1994); Faaborg et al. (1995); and Bender, Contreras, and Fahrig (1998) reviewed the consequences of habitat fragmentation to populations of birds and other species. For examples of some negative impacts of habitat fragmentation to woodlands birds of the eastern United States, see Robbins, Dawson, and Dowell 1989; Robinson et al. 1995; and Robinson 1998. Herkert (1994) and Vickery, Hunter, and Melvin (1994) found similar effects among grassland birds in the Midwest and Northeast, as did Bolger, Scott, and Rotenberry (1997) and Powell and Collier (1998) for birds in chaparral and grassland habitats in southern California. T. E. Martin (1995) synthesized information on the importance of nest predation to the evolutionary ecology of birds, while the study by Keyser, Hill, and Soehren (1998) is a recent example showing the positive relationship between avian nest predation and habitat fragmentation.

For recent examples of the impacts of urbanization and outdoor recreation on wildlife populations, see Knight and Gutzwiller 1995; Blair 1996; and Germaine et al. 1998. Wilcove 1985; Potter 1991; Andrén 1992; Mitchell and Beck 1992; and Haskell 1995 include examples of the impacts of human commensal and domestic predators, including birds in the family Corvidae, on avian populations. Naeser and St. John (1998) reviewed water use in relation to present and future suburbanization of the Sonoita Valley.

A "cows vs. condos" debate centers on the relative impacts of suburban and second-home development versus livestock grazing on vegetation and wildlife in the American West (e.g., Wuerthner 1994; J. H. Brown and McDonald 1995). There seems little doubt that the average grazed acre can support more native biological diversity than a series of suburban backyards. However, the majority of western livestock grazing takes place on public (usually federal) rangelands that cannot be developed, barring some drastic change in land disposition.

CHAPTER 10. THE NEST BOX EXPERIMENT

The community theory of ecological organization was advanced first by Clements (1916), while the individualistic view was proposed by Gleason (1926). For reviews from a plant ecological perspective, see Whittaker 1975 and Callaway 1997. Among animal ecologists, much of the debate has centered on the importance (or lack of it) of interspecific competition among species. For a diversity of views on this subject, see Schoener 1982; D. R. Strong 1983; Andrewartha and Birch 1984; J. H. Brown et al. 1986; Diamond and Case 1986; and Keddy 1989. Schoener (1983), Connell (1983), and Goldberg and Barton (1992) reviewed and synthesized results of field experiments testing the importance of interspecific competition in structuring biological communities.

Grinnell (1917, 1928) and MacArthur (1958, 1972) were especially prominent avian ecologists who wrote about the importance of competition among birds. Wiens (1989) provides an excellent synthesis of this subject, including the work of his own research team suggesting that competition may not always play an important role in determining avian habitat selection, population dynamics, or community structure (Wiens and Rotenberry 1981a,b; Wiens 1983; Rotenberry and Wiens 1998).

Nest box experiments similar to the one we conducted in southeastern Arizona (C. E. Bock et al. 1992) have been performed in ponderosa pine forests of northern Arizona (Brawn, Boecklen, and Balda 1987; Brawn and Balda 1988), in mixed coniferous forests of the Colorado Front Range (C. E. Bock and Fleck 1995), and at various locations in Britain and Europe (Enemar and Sjostrand 1972; Hogstad 1975; East and Perrins 1988; Mönkkönen, Helle, and Soppela 1990). Nest sites usually are limiting to cavity-nesting birds (Brush 1983; Gustafsson 1988; but see Waters, Noon, and Verner 1990), so that nest box additions usually have increased their local densities. However, we are aware of only one other case besides ours in Arizona where nest box addition also resulted in declines of open cup–nesting species on experimental plots (Hogstad 1975).

CHAPTER 11. PLAINS LOVEGRASS

Plains lovegrass (*Eragrostis intermedia*) is distributed from Georgia to Arizona and south to Central America (Gould 1951). It is widely known to be sensitive to livestock grazing (e.g., Humphrey 1970). At the Research Ranch, we have documented its post-grazing increases, as well as the effects of fire and drought (Brady et al. 1989; C. E. Bock and J. H. Bock 1993; C. E. Bock et al. 1995).

The independent and interactive responses of plains lovegrass to drought, fire, and grazing typify the inherently dynamic nature of grassland ecosystems as a whole, although the relative importance of these three forces varies among different species and grassland communities (Mack and Thompson 1982; Milchunas, Sala, and Lauenroth 1988; Belsky 1992; Seastedt and Knapp 1993). Because grasslands are naturally variable on large temporal and spatial scales, effective management and conservation are especially challenging (e.g., S. L. Collins and Glenn 1995). Four recent and important books on this subject are *Secondary Succession and the Evaluation of Rangeland Condition,* edited by Lauenroth and Laycock (1989); *The Changing Prairie,* edited by Joern and Keeler (1995); *Prairie Conservation,* edited by Samson and Knopf (1996); and *Ecology and Economics of the Great Plains,* by Licht (1997).

For perspectives on antienvironmental aspects of the wise use movement, see Helvarg 1994; Echeverria and Eby 1995; and Ehrlich and Ehrlich 1996.

CHAPTER 12. COTTON RATS AND REAL DOGS

For general information about southwestern rodents and their predators, including impacts of livestock grazing and predator control, see Hoffmeister 1986; C. Jones and Manning 1996; Fagerstone and Ramey 1996; and Andelt 1996. In an exemplary long-term study, James Brown and his colleagues have studied population dynamics, species interactions, and ecological effects of the rodent assemblage in a Chihuahuan Desert site near Portal, Arizona (e.g., J. H. Brown et al. 1986; J. H. Brown and Zeng 1989; J. H. Brown and Heske 1990; Valone and Brown 1995).

Merriam's kangaroo rats are well known to prefer desert grasslands with relatively large amounts of bare ground (Reynolds 1958; Rosenzweig and Winakur 1969), whereas rodents such as western harvest mice, cotton rats, and hispid pocket mice usually are more common in areas with higher amounts of grass cover (e.g., Whitford 1976; Fagerstone and Ramey 1996; Hayward, Heske, and Painter 1997). Results from rodent live-trapping on the Research Ranch and adjacent grazed lands were reported in J. H. Bock, Bock, and McKnight 1976; C. E. Bock and J. H. Bock 1978; and C. E. Bock et al. 1984, 1986. For additional studies on rodent responses to grazing in arid grasslands, see Heske and Campbell 1991 and Hayward, Heske, and Painter 1997.

For general information about the ecology and behavior of coyotes, and examples of recent studies, see Bekoff 1978; Leydet 1988; Meinzer 1995; and Windberg, Ebbert, and Kelly 1997. D. E. Brown (1983) and Burbank (1990) described the history of the Mexican wolf in the Southwest. Ortega (1987a) reported on coyote food habits at the Research Ranch.

Sih et al. (1985) reviewed literature on the overall importance of predation in biological communities, as evidenced by the results of field experiments. Data from predator exclosures on the Research Ranch suggest that desert grassland rodents, especially cotton rats, can be limited by predation (C. E. Bock and J. H. Bock 1994). Field experiments and observations from other communities have shown similar results (e.g., Schnell 1968; Wiegert 1972; Desy and Batzli 1989; Longland and Price 1991; Kotler, Brown, and Mitchell 1994; Batzli and Lesieutre 1995; Meserve et al. 1996; Stapp 1997), suggesting that predators can influence the distribution, abundance, and behavior of rodents under many circumstances.

CHAPTER 13. ALIENS

Two now-classic works on the spread of alien species are *The Ecology of Invasions by Plants and Animals* by Elton (1958) and *The Genetics of Colonizing Species*, edited by Baker and Stebbins (1965). Recent general reviews and synthetic articles on exotic species include Mooney and Drake 1986; Drake et al. 1989; Hengeveld 1989; Pysek et al. 1995; Luken and Thieret 1997; Schmitz and Simberloff 1997; Devine 1998; and Richards, Chambers, and Ross 1998. The history and consequences of invasions by exotic grasses in particular were reviewed by D'Antonio and Vitousek (1992); Lonsdale (1994); and Burke and Grime (1996). Brock (1998); Tellman (1998); and J. H. Bock and C. E. Bock (1995, 2000) reviewed the spread of aliens specifically into arid southwestern grasslands.

Berger (1993) described the problems associated with exotic species in ecological restoration of native ecosystems. Bengtsson, Jones, and Setala (1997) and Hooper and Vitousek (1998) considered the question of whether native biodiversity is important to ecosystem services, and the consequences of replacing species-rich endemic communities with species-poor assemblages dominated by exotics. For general syntheses on various aspects of species diversity, see Huston 1994 and Rosenzweig 1995. Various chapters in Daily 1997 provide contemporary reviews of the value of ecosystem services in general. For recent views on the aesthetic, spiritual, and moral aspects to the conservation of biological diversity, including grasslands, see Rolston 1994; Kim and Weaver 1994; Manning 1995; and Snyder 1995.

For a history of the spread of Lehmann lovegrass into southeastern Arizona and reviews of its ecological requirements and responsiveness to fire, grazing, and reseeding with native grasses, see Cox and Ruyle 1986; Ruyle, Roundy, and Cox 1988; Cox, Ruyle, and Roundy 1990; Anable, McLaran, and Ruyle 1992; Robinett 1992; Roundy and Biedenbender 1995; and Biedenbender and Roundy 1996.

Freeman (1979) described the introduction of Lehmann lovegrass into the Sonoita Valley in 1940. We compared the biodiversity of native grasslands versus areas on the Research Ranch dominated by Lehmann and Boer lovegrass (C. E. Bock et al. 1986) and the consequences of wildfire in both sorts of communities (C. E. Bock and J. H. Bock 1992; J. H. Bock and C. E. Bock 1992b). Whitford (1997) compared abundances of birds and mammals between native and Lehmann lovegrass stands elsewhere in southeastern Arizona.

For the history, ecology, and environmental consequences of tamarisk invasions into southwestern riparian ecosystems, see Busch and Smith 1995; Turner, Bowers, and Burgess 1995; Sala, Smith, and Devitt 1996; Cleverly et al. 1997; and Tellman 1998.

CHAPTER 14. THE WORLD IS FULL OF LIFE

There is an extensive literature about human population growth, and its relationships to the environment, conservation, economics, and social justice. For discussions of human population growth and its historical and projected environmental impacts, see especially Ehrlich and Ehrlich 1990; Cohen 1995; Grant 1996; Ehrlich 1997; Eldredge 1997; and McKibben 1998. Dowie (1995); Bullard (1996); Vandermeer (1996); Low and Gleeson (1998); and Wenz (1998) consider the social, political, economic, and equity aspects of the issue. Rush Limbaugh wrote *The Way Things Ought to Be* (1992). Norman Maclean's observation about the nature of man, which he attributed to his father, is from *A River Runs through It* (1976).

LITERATURE CITED

Abbey, E. 1975. *The Monkey Wrench Gang.* Philadelphia: J. B. Lippincott.

American Ornithologists' Union. 1983. *Checklist of North American Birds.* 6th ed. Washington, D.C.: American Ornithologists' Union.

Anable, M. E., M. P. McClaran, and G. B. Ruyle. 1992. Spread of introduced Lehmann lovegrass (*Eragrostis lehmanniana* Nees) in southern Arizona, USA. *Biological Conservation* 61:181–188.

Andelt, W. F. 1996. Carnivores. In *Rangeland Wildlife,* ed. P. R. Krausman, 133–155. Denver: Society for Range Management.

Anderson, B. W., and R. D. Ohmart. 1977. Vegetation structure and bird use in the Lower Colorado River Valley. In *Importance, Preservation, and Management of Riparian Habitats: A Symposium,* R. R. Johnson and D. A. Jones, technical coordinators, 23–34. USDA Forest Service, General Technical Report RM-43. Fort Collins, Colo.: Rocky Mountain Forest and Range Experiment Station.

Anderson, R. C. 1982. An evolutionary model summarizing the roles of fire, climate, and grazing animals in the origin and maintenance of grasslands: an end paper. In *Grasses and Grasslands: Systematics and Ecology,* ed. J. R. Estes, R. J. Tyrl, and J. N. Brunken, 297–308. Norman: University of Oklahoma Press.

Andrén, H. 1992. Corvid density and nest predation in relation to forest fragmentation: a landscape perspective. *Ecology* 73:794–804.

———. 1994. Effects of habitat fragmentation on birds and mammals in landscapes with different proportions of suitable habitat: a review. *Oikos* 71:355–366.

Andrewartha, H. G., and L. C. Birch. 1984. *The Ecological Web.* Chicago: University of Chicago Press.

Arcese, P., and A. R. E. Sinclair. 1997. The role of protected areas as ecological baselines. *Journal of Wildlife Management* 61:587–602.

Archer, S. 1994. Woody plant encroachment into southwestern grasslands and savannas: rates, patterns, and proximate causes. In *Ecological Implications of Livestock Herbivory in the West,* ed. M. Vavra, W. A. Laycock, and R. D. Pieper, 13–68. Denver: Society for Range Management.

Archer, S., D. S. Schimel, and E. A. Holland. 1995. Mechanisms of shrubland expansion: land use, climate or CO_2? *Climatic Change* 29:91–99.

Axelrod, D. I. 1985. The rise of the grassland biome, central North America. *Botanical Review* 51:163–201.

Bahre, C. J. 1977. Land-use history of the Research Ranch, Elgin, Arizona. *Journal of the Arizona Academy of Science* 12 (suppl. 2): 1–32.

———. 1985. Wildfire in southeastern Arizona between 1859 and 1890. *Desert Plants* 7:190–194.

———. 1991. *A Legacy of Change: Historic Human Impact on Vegetation of the Arizona Borderlands.* Tucson: University of Arizona Press.

———. 1995. Human impacts on the grasslands of southeastern Arizona. In *The Desert Grassland,* ed. M. P. McClaran and T. R. Van Devender, 230–264. Tucson: University of Arizona Press.

———. 1998. Late nineteenth century human impacts on the woodlands and forests of southeastern Arizona's sky islands. *Desert Plants* 14:8–21.

Bahre, C. J., and M. L. Shelton. 1993. Historic vegetation change, mesquite increases, and climate in southeastern Arizona. *Journal of Biogeography* 20:489–504.

———. 1996. Rangeland destruction: cattle and drought in southeastern Arizona at the turn of the century. *Journal of the Southwest* 38:1–22.

Baker, H. G., and G. L. Stebbins, eds. 1965. *The Genetics of Colonizing Species.* New York: Academic Press.

Ballinger, R. E., and J. D. Congdon. 1996. Status of the bunch grass lizard, *Sceloporus scalaris,* in the Chiricahua Mountains of southeastern Arizona. *Bulletin of the Maryland Herpetological Society* 32:67–69.

Barstad, J. F. 1981. Factors controlling plant distribution in a riparian deciduous forest in southeastern Arizona. M.A. thesis, Arizona State University.

Batzli, G. O., and C. Lesieutre. 1995. Community organization of arvicoline rodents in northern Alaska. *Oikos* 72:88–98.

Bekoff, M., ed. 1978. *Coyotes: Biology, Behavior, and Management.* New York: Academic Press.

Belovsky, G. E., and J. B. Slade. 1993. The role of vertebrate and invertebrate predators in a grasshopper community. *Oikos* 68:193–201.

Belsky, J. 1992. Effects of grazing, competition, disturbance, and fire on species composition and diversity in grassland communities. *Journal of Vegetation Science* 3:187–200.

Bender, D. J., T. A. Contreras, and L. Fahrig. 1998. Habitat loss and population decline: a meta-analysis of the patch size effect. *Ecology* 79:517–533.

Bengtsson, J., H. Jones, and H. Setala. 1997. The value of biodiversity. *Trends in Ecology and Evolution* 12:334–336.

Berger, J.J. 1993. Ecological restoration and nonindigenous plant species: a review. *Restoration Ecology* 1:74–82.
Betancourt, J., T.R. Van Devender, and P.S. Martin. 1990. *Packrat Middens: The Last 40,000 Years of Biotic Change.* Tucson: University of Arizona Press.
Biedenbender, S.H., and B.A. Roundy. 1996. Establishment of native semidesert grasses into existing stands of *Eragrostis lehmanniana* in southeastern Arizona. *Restoration Ecology* 4:155–162.
Bissonette, J.A. 1982. *Social Behavior and Ecology of the Collared Peccary in Big Bend National Park.* U.S. National Park Service, Science Monograph No. 16. Washington, D.C.: USNPS, Department of Interior.
Blair, R.B. 1996. Land use and avian species diversity along an urban gradient. *Ecological Applications* 6:506–519.
Blaustein, A.R., and D.B. Wake. 1995. The puzzle of declining amphibian populations. *Scientific American* 272 (4): 52–57.
Bock, C.E., and J.H. Bock. 1978. Responses of birds, small mammals, and vegetation to burning of sacaton grassland in southeastern Arizona. *Journal of Range Management* 31:289–300.
———. 1979. Relationship of the collared peccary to sacaton grassland. *Journal of Wildlife Management* 43:813–816.
———. 1984. Importance of sycamores to riparian birds in southeastern Arizona. *Journal of Field Ornithology* 55:97–103.
———. 1988. Grassland birds in southeastern Arizona: impacts of fire, grazing, and alien vegetation. In *Ecology and Conservation of Grassland Birds,* ed. P. Goriup, 43–58. Technical Publication No. 7. Cambridge, Eng.: International Council for Bird Preservation.
———. 1991. Response of grasshoppers (Orthoptera: Acrididae) to wildfire in a southeastern Arizona grassland. *American Midland Naturalist* 125:162–167.
———. 1992. Response of birds to wildfire in native versus exotic Arizona grassland. *Southwestern Naturalist* 37:73–81.
———. 1993. Cover of perennial grasses in southeastern Arizona in relation to livestock grazing. *Conservation Biology* 7:371–377.
———. 1994. Effects of predator exclusion on rodent abundance in an Arizona semidesert grassland. *Southwestern Naturalist* 39:208–210.
———. 1997. Shrub densities in relation to fire, livestock grazing, and precipitation in an Arizona desert grassland. *Southwestern Naturalist* 42:188–193.
———. 1999. Response of birds to drought and short-duration grazing in southeastern Arizona. *Conservation Biology* 13:1117–1123.
Bock, C.E., J.H. Bock, and M.C. Grant. 1992. Effects of bird predation on grasshopper densities in an Arizona grassland. *Ecology* 73:1706–1717.
Bock, C.E., J.H. Bock, M.C. Grant, and T.R. Seastedt. 1995. Effects of fire on abundance of plains lovegrass (*Eragrostis intermedia*) in a semiarid grassland in southeastern Arizona. *Journal of Vegetation Science* 6:325–328.
Bock, C.E., J.H. Bock, K.L. Jepson, and J.C. Ortega. 1986. Ecological effects of planting African lovegrasses in Arizona. *National Geographic Research* 2:456–463.

Bock, C. E., J. H. Bock, W. R. Kenney, and V. M. Hawthorne. 1984. Response of birds, rodents, and vegetation to livestock exclosure in a semidesert grassland site. *Journal of Range Management* 37:239–242.

Bock, C. E., A. Cruz Jr., M. C. Grant, C. S. Aid, and T. R. Strong. 1992. Field experimental evidence for diffuse competition among southwestern riparian birds. *American Naturalist* 140:815–828.

Bock, C. E., and D. C. Fleck. 1995. Avian response to nest box addition in two forests of the Colorado Front Range. *Journal of Field Ornithology* 66:352–362.

Bock, C. E., H. M. Smith, and J. H. Bock. 1990. The effect of livestock grazing upon abundance of the lizard, *Sceloporus scalaris,* in southeastern Arizona. *Journal of Herpetology* 24:445–446.

Bock, C. E., and B. Webb. 1984. Birds as grazing indicator species in southeastern Arizona. *Journal of Wildlife Management* 48:1045–1049.

Bock, J. H., and C. E. Bock. 1985. Patterns of reproduction in Wright's sycamore. In *Riparian Ecosystems and Their Management,* R. R. Johnson, C. D. Ziebell, D. R. Patton, P. F. Ffolliott, and R. H. Hamre, technical coordinators, 493–494. USDA Forest Service, General Technical Report RM-120. Fort Collins, Colo.: Rocky Mountain Forest and Range Experiment Station.

———. 1986. Habitat relationships of some native perennial grasses in southeastern Arizona. *Desert Plants* 8:3–14.

———. 1989. Factors limiting sexual reproduction in *Platanus wrightii* in southeastern Arizona. *Aliso* 12:295–301.

———. 1992a. Short-term reductions in plant densities following fire in an ungrazed semidesert shrub-grassland. *Southwestern Naturalist* 37:49–53.

———. 1992b. Vegetation responses to wildfire in native vs. exotic Arizona grassland. *Journal of Vegetation Science* 3:439–446.

———. 1995. The challenges of grassland conservation. In *The Changing Prairie: North American Grasslands,* ed. A. Joern and K. H. Keeler, 199–222. New York: Oxford University Press.

———. 2000. Exotic species in grasslands. In *Invasive Exotic Species in the Sonoran Desert Region,* ed. B. Tellman. In press.

Bock, J. H., C. E. Bock, and J. R. McKnight. 1976. A study of the effects of grassland fires at the Research Ranch in southeastern Arizona. *Journal of the Arizona Academy of Science* 11:49–57.

Bolger, D. T., T. A. Scott, and J. T. Rotenberry. 1997. Breeding bird abundance in an urbanizing landscape in coastal southern California. *Conservation Biology* 11:406–421.

Bonham, C. D. 1972. *Ecological Inventory and Data Storage Retrieval System for the Research Ranch.* Range Science Department, Science Series No. 14. Fort Collins: Colorado State University.

Bowers, J. E., and S. P. McLaughlin. 1996. Flora of the Huachuca Mountains, a botanically rich and historically significant sky island in Cochise County, Arizona. *Journal of the Arizona-Nevada Academy of Science* 29:65–107.

Brady, W. W., M. R. Stromberg, E. F. Aldon, C. D. Bonham, and S. H. Henry. 1989. Response of a semidesert grassland to sixteen years of rest from grazing. *Journal of Range Management* 42:284–288.

Bragg, T. B. 1995. The physical environment of Great Plains grasslands. In *The Changing Prairie: North American Grasslands,* ed. A. Joern and K. H. Keeler, 49–81. New York: Oxford University Press.

Bragg, T. B., and L. C. Hulbert. 1976. Woody plant invasion of unburned Kansas bluestem prairie. *Journal of Range Management* 29:19–24.

Brawn, J. D., and R. P. Balda. 1988. Population biology of cavity nesters in northern Arizona: do nest sites limit breeding densities? *Condor* 90:61–71.

Brawn, J. D., W. J. Boecklen, and R. P. Balda. 1987. Investigations of density interactions among breeding birds in ponderosa pine forests: correlative and experimental evidence. *Oecologia* 72:348–357.

Brawn, J. D., and S. K. Robinson. 1996. Source-sink dynamics may complicate the interpretation of long-term census data. *Ecology* 77:3–12.

Brock, J. H. 1998. Ecological characteristics of invasive alien plants. In *The Future of Arid Grasslands: Identifying Issues, Seeking Solutions,* ed. B. Tellman, D. M. Finch, C. Edminster, and R. Hamre, 137–143. Proceedings RMRS-P-3. Fort Collins, Colo.: USDA Forest Service, Rocky Mountain Research Station.

Brown, D. E., ed. 1982. Biotic communities of the American Southwest—United States and Mexico. *Desert Plants* 4:1–342.

———, ed. 1983. *The Wolf in the Southwest.* Tucson: University of Arizona Press.

Brown, D. E., and R. Davis. 1998. Terrestrial bird and mammal distribution changes in the American Southwest, 1890–1990. In *The Future of Arid Grasslands: Identifying Issues, Seeking Solutions,* ed. B. Tellman, D. M. Finch, C. Edminster, and R. Hamre, 47–64. Proceedings RMRS-P-3. Fort Collins, Colo.: USDA Forest Service, Rocky Mountain Research Station.

Brown, J. H. 1995. *Macroecology.* Chicago: University of Chicago Press.

Brown, J. H., D. W. Davidson, J. C. Munger, and R. S. Inouye. 1986. Experimental community ecology: the desert granivore system. In *Community Ecology,* ed. J. Diamond and T. J. Case, 41–61. New York: Harper & Row.

Brown, J. H., and E. J. Heske. 1990. Control of a desert-grassland transition by a keystone rodent guild. *Science* 250:1705–1707.

Brown, J. H., and W. McDonald. 1995. Livestock grazing and conservation on southwestern rangelands. *Conservation Biology* 9:1644–1647.

Brown, J. H., and Z. Zeng. 1989. Comparative population ecology of eleven species of rodents in the Chihuahuan Desert. *Ecology* 70:1507–1525.

Brown, J. L. 1994. Mexican jay. In *The Birds of North America,* No. 118, ed. A. Poole and F. Gill. Philadelphia: Academy of Natural Sciences; Washington, D.C.: American Ornithologists' Union.

Brown, J. L., and E. R. Brown. 1990. Mexican jays: uncooperative breeding. In *Cooperative Breeding in Birds,* ed. P. B. Stacey and W. D. Koenig, 239–266. Cambridge, Eng.: Cambridge University Press.

Brown, J. L., E. R. Brown, J. Sedransk, and S. Ritter. 1997. Dominance, age, and reproductive success in a complex society: a long-term study of the Mexican jay. *Auk* 114:279–286.

Brush, T. 1983. Cavity use by secondary cavity-nesting birds and response to manipulations. *Condor* 85:461–466.

Brussard, P. F., D. D. Murphy, and C. R. Tracy. 1994. Cattle and conservation biology—another view. *Conservation Biology* 8:919–921.

Buffington, L. C., and C. H. Herbel. 1965. Vegetational changes on a semidesert grassland range from 1858 to 1963. *Ecological Monographs* 35:139–164.

Bullard, R. D., ed. 1996. *Unequal Protection: Environmental Justice and Communities of Color.* San Francisco: Sierra Club Books.

Burbank, J. C. 1990. *Vanishing Lobo: The Mexican Wolf and the Southwest.* Boulder, Colo.: Johnson Books.

Burgess, T. L. 1995. Desert grassland, mixed shrub savanna, shrub steppe, or semidesert scrub? The dilemma of coexisting growth forms. In *The Desert Grassland,* ed. M. P. McClaran and T. R. Van Devender, 31–67. Tucson: University of Arizona Press.

Burke, M. J., and J. P. Grime. 1996. An experimental study of plant community invasibility. *Ecology* 77:776–790.

Busch, D. E., and S. D. Smith. 1995. Mechanisms associated with decline of woody species in riparian ecosystems of the southwestern U.S. *Ecological Monographs* 65:347–370.

Byers, J. A., and M. Bekoff. 1981. Social spacing and cooperative behavior of collared peccaries. *Journal of Mammalogy* 62:767–785.

Callaway, R. M. 1997. Positive interactions in plant communities and the individualistic-continuum concept. *Oecologia* 112:143–149.

Carothers, S. W., R. R. Johnson, and S. W. Aitchison. 1974. Population and social organization of southwestern riparian birds. *American Zoologist* 14:97–108.

Chew, R. M. 1982. Changes in herbaceous and suffrutescent perennials in grazed and ungrazed desertified grassland in southeastern Arizona, 1958–1978. *American Midland Naturalist* 108:159–169.

Clarke, A. C. 1968. *2001: A Space Odyssey.* New York: New American Library.

Clements, F. E. 1916. *Plant Succession: Analysis of the Development of Vegetation.* Publication No. 242. Washington, D.C.: Carnegie Institute.

———. 1920. *Plant Indicators.* Publication No. 290. Washington, D.C.: Carnegie Institute.

Cleverly, J. R., S. D. Smith, A. Sala, and D. A. Devitt. 1997. Invasive capacity of *Tamarix ramosissima* in a Mojave Desert floodplain: the role of drought. *Oecologia* 111:12–18.

Cohen, J. E. 1995. *How Many People Can the Earth Support?* New York: W. W. Norton.

Collins, J. T. 1997. Standard common and current scientific names for North American amphibians and reptiles. Herpetological Circular No. 25. Lawrence, Kans.: Society for the Study of Amphibians and Reptiles.

Collins, S. L., and S. M. Glenn. 1995. Grassland ecosystem and landscape dynamics. In *The Changing Prairie: North American Grasslands,* ed. A. Joern and K. H. Keeler, 128–156. New York: Oxford University Press.
Connell, J. H. 1983. On the prevalence and relative importance of interspecific competition: evidence from field experiments. *American Naturalist* 122:661–696.
Connor, E. F., and D. Simberloff. 1986. Competition, scientific method, and null models in ecology. *American Scientist* 74:155–162.
Cox, J. R., F. A. Ibarra F., and M. H. Martin R. 1990. Fire effects on grasses in semiarid deserts. In *Effects of Fire Management of Southwestern Natural Resources,* ed. J. Krammes, 43–49. USDA Forest Service, General Technical Publication RM-91. Fort Collins, Colo.: Rocky Mountain Forest and Range Experiment Station.
Cox, J. R., and G. B. Ruyle. 1986. Influence of climatic and edaphic factors on the distribution of *Eragrostis lehmanniana* Nees in Arizona, USA. *Journal of the Grassland Society of Southern Africa* 3:25–29.
Cox, J. R., G. B. Ruyle, and B. A. Roundy. 1990. Lehmann lovegrass in southeastern Arizona: biomass production and disappearance. *Journal of Range Management* 43:367–372.
Creer, D. A., K. M. Kjer, D. L. Simmons, and J. W. Sites Jr. 1997. Phylogenetic relationships of the *Sceloporus scalaris* species group (Squamata). *Journal of Herpetology* 31:353–364.
Cross, A. F. 1991. Vegetation of two southeastern Arizona desert marshes. *Madroño* 38:185–194.
Daily, G. C., ed. 1997. *Nature's Services: Societal Dependence on Natural Ecosystems.* Washington, D.C.: Island Press.
D'Antonio, C. M., and P. M. Vitousek. 1992. Biological invasions by exotic grasses, the grass/fire cycle, and global change. *Annual Review of Ecology and Systematics* 23:63–87.
Davis, S. K., and S. G. Sealy. 1998. Nesting biology of the Baird's sparrow in southwestern Manitoba. *Wilson Bulletin* 110:262–270.
Davis, W. A., and S. M. Russell. 1995. *Finding Birds in Southeast Arizona.* Tucson: Tucson Audubon Society.
DeBano, L. F., D. G. Neary, and P. F. Ffolliott. 1998. *Fire's Effects on Ecosystems.* New York: John Wiley.
Desy, E. A., and G. O. Batzli. 1989. Effects of food availability and predation on prairie vole demography: a field experiment. *Ecology* 70:411–421.
Devine, R. 1998. *Alien Invasion.* Washington, D.C.: National Geographic Society.
Diamond, J., and T. J. Case, eds. 1986. *Community Ecology.* New York: Harper & Row.
Diffendorfer, J. E. 1998. Testing models of source-sink dynamics and balanced dispersal. *Oikos* 81:417–433.
Dobkin, D. S., A. C. Rich, and W. H. Pyle. 1998. Habitat and avifaunal recovery from livestock grazing in a riparian meadow system of the northwestern Great Basin. *Conservation Biology* 12:209–221.

Dobyns, H. F. 1981. *From Fire to Flood: Historic Human Destruction of Sonoran Desert Riverine Oases.* Anthropology Paper No. 20. Socorro, N.M.: Ballena Press.

Dowie, M. 1995. *Losing Ground: American Environmentalism at the Close of the Twentieth Century.* Cambridge, Mass.: MIT Press.

Drake, J., F. Di Castri, R. Groves, F. Kruger, H. Mooney, M. Rejmanek, and M. Williamson, eds. 1989. *Biological Invasions: A Global Perspective.* Chichester, Eng.: John Wiley.

Dufrene, M., and P. Legendre. 1997. Species assemblages and indicator species: the need for a flexible asymmetrical approach. *Ecological Monographs* 67:345–366.

Dunning, J. B., Jr., R. K. Bowers, S. J. Suter, and C. E. Bock. 1999. Cassin's sparrow. In *The Birds of North America,* ed. A. Poole and F. Gill. Philadelphia: Academy of Natural Sciences; Washington, D.C.: American Ornithologists' Union, in press.

Dunning, J. B., Jr., and J. H. Brown. 1982. Summer rainfall and winter sparrow densities: a test of the food limitation hypothesis. *Auk* 99:123–129.

Dunning, J. B., Jr., D. J. Stewart, B. J. Danielson, B. R. Noon, T. L. Root, R. H. Lamberson, and E. E. Stevens. 1995. Spatially explicit population models: current forms and future uses. *Ecological Applications* 5:3–11.

East, M. L., and C. M. Perrins. 1988. The effect of nestboxes on breeding populations of birds in broadleaved temperate woodlands. *Ibis* 130:393–401.

Echeverria, J. D., and R. B. Eby, eds. 1995. *Let the People Judge: Wise Use and the Private Property Rights Movement.* Washington, D.C.: Island Press.

Ehrlich, P. R. 1997. *A World of Wounds: Ecologists and the Human Dilemma.* Oldendorf/Luhe, Ger.: Ecology Institute.

Ehrlich, P. R., and A. H. Ehrlich. 1990. *The Population Explosion.* New York: Simon & Schuster.

———. 1996. *Betrayal of Science and Reason: How Anti-environmental Rhetoric Threatens Our Future.* Washington, D.C.: Island Press.

Eldredge, N. 1997. *Dominion.* Berkeley: University of California Press.

Elton, C. S. 1958. *The Ecology of Invasions by Animals and Plants.* London: Methuen.

Enemar, A., and B. Sjostrand. 1972. Effects of the introduction of pied flycatchers on the composition of a passerine bird community. *Ornis Scandinavica* 3:79–89.

Faaborg, J., M. Brittingham, T. Donovan, and J. Blake. 1995. Habitat fragmentation in the temperate zone. In *Ecology and Management of Neotropical Migratory Birds,* ed. T. E. Martin and D. M. Finch, 357–380. New York: Oxford University Press.

Fagerstone, K. A., and C. A. Ramey. 1996. Rodents and lagomorphs. In *Rangeland Wildlife,* ed. P. R. Krausman, 83–132. Denver: Society for Range Management.

Ferguson, D., and N. Ferguson. 1983. *Sacred Cows at the Public Trough.* Bend, Oreg.: Maverick Publications.

Fleck, D. C., and D. F. Tomback. 1996. Tannin and protein in the diet of a food-hoarding granivore, the western scrub-jay. *Condor* 98:474–482.

Fleck, D. C., and G. E. Woolfenden. 1997. Can acorn tannin predict scrub-jay caching behavior? *Journal of Chemical Ecology* 23:793–806.

Fleischner, T. L. 1994. Ecological costs of livestock grazing in western North America. *Conservation Biology* 8:629–644.
———. 1996. Reply to Brown and McDonald. *Conservation Biology* 10:927–929.
Floyd, T. 1996. Top-down impacts on creosotebush herbivores in a spatially and temporally complex environment. *Ecology* 77:1544–1555.
Forman, R. T. T., and M. Gordon. 1986. *Landscape Ecology.* New York: John Wiley.
Freeman, D. 1979. Lehmann lovegrass. *Rangelands* 1:162–163.
Freemark, K. E., J. B. Dunning Jr., S. J. Hejl, and J. R. Probst. 1995. A landscape ecology perspective for research, conservation, and management. In *Ecology and Management of Neotropical Migratory Birds,* ed. T. E. Martin and D. M. Finch, 381–427. New York: Oxford University Press.
Gehlbach, F. R. 1981. *Mountain Islands and Desert Seas: A Natural History of the U.S.-Mexico Borderlands.* College Station: Texas A & M University Press.
Germaine, S. S., S. S. Rosenstock, R. E. Schweinsburg, and W. S. Richardson. 1998. Relationships among breeding birds, habitat, and residential development in greater Tucson, Arizona. *Ecological Applications* 8:680–691.
Gibson, D. J., and L. C. Hulbert. 1987. Effects of fire, topography, and year-to-year climatic variation on species composition in tallgrass prairie. *Vegetatio* 72:175–185.
Gleason, H. A. 1926. The individualistic concept of the plant association. *Bulletin of the Torrey Botanical Club* 53:7–26.
Goldberg, D. E., and A. M. Barton. 1992. Patterns and consequences of interspecific competition in natural communities: a review of field experiments with plants. *American Naturalist* 139:771–801.
Golluscio, R. A., O. E. Sala, and W. K. Lauenroth. 1998. Differential use of large summer rainfall events by shrubs and grasses: a manipulative experiment in the Patagonia steppe. *Oecologia* 115:17–25.
Goriup, P. D., ed. 1988. *Ecology and Conservation of Grassland Birds.* Technical Publication No. 7. Cambridge, Eng.: International Council for Bird Preservation.
Gould, F. W. 1951. *Grasses of Southwestern United States.* Tucson: University of Arizona Press.
Grant, L. 1996. *Juggernaut: Growth on a Finite Planet.* Santa Ana, Calif.: Seven Locks Press.
Grinnell, J. 1917. The niche-relationships of the California thrasher. *Auk* 34:427–433.
———. 1922. The role of the "accidental." *Auk* 39:373–380.
———. 1928. The presence and absence of animals. *University of California Chronicle* 30:429–450.
Guillette, L. J., Jr., R. E. Jones, K. T. Fitzgerald, and H. M. Smith. 1980. Evolution of viviparity in the lizard genus *Sceloporus. Herpetologica* 36:201–215.
Guo, Q., and J. H. Brown. 1997. Interactions between winter and summer annuals in the Chihuahuan Desert. *Oecologia* 111:123–128.
Gustafsson, L. 1988. Inter- and intraspecific competition for nest holes in a population of the collared flycatcher *Fidecula albicollis. Ibis* 130:11–16.

Hall, H. M., and J. Grinnell. 1919. Life-zone indicators in California. *Proceedings of the California Academy of Science,* 4th ser., 9:37–67.
Hanski, I. A., and M. E. Gilpin, eds. 1997. *Metapopulation Biology: Ecology, Genetics, and Evolution.* San Diego: Academic Press.
Haskell, D. G. 1995. A reevaluation of the effects of forest fragmentation on rates of bird-nest predation. *Conservation Biology* 9:1316–1318.
Hastings, R. M., and R. M. Turner. 1965. *The Changing Mile: An Ecological Study of Vegetation Change with Time in the Lower Mile of an Arid and Semiarid Region.* Tucson: University of Arizona Press.
Hayward, B., E. J. Heske, and C. W. Painter. 1997. Effects of livestock grazing on small mammals at a desert cienega. *Journal of Wildlife Management* 61:123–129.
Heitschmidt, R. K., S. L. Dowhower, and J. W. Walker. 1987. 14- vs. 42-paddock rotational grazing: aboveground biomass dynamics, forage production, and harvest efficiency. *Journal of Range Management* 40:216–223.
Hellgren, E. C., D. R. Synatzske, P. W. Oldenburg, and F. S. Guthery. 1995. Demography of a collared peccary population in south Texas. *Journal of Wildlife Management* 59:153–163.
Helvarg, D. 1994. *The War against the Greens: The "Wise-Use" Movement, the New Right, and Anti-environmental Violence.* San Francisco: Sierra Club Books.
Hengeveld, R. 1989. *Dynamics of Biological Invasions.* London: Chapman & Hall.
Herkert, J. R. 1994. The effects of habitat fragmentation on midwestern grassland bird communities. *Ecological Applications* 4:461–471.
Heske, E. J., and M. Campbell. 1991. Effects of an eleven-year livestock exclosure on rodent and ant numbers in the Chihuahuan Desert, southeastern Arizona. *Southwestern Naturalist* 36:89–93.
Hoffmeister, D. F. 1986. *Mammals of Arizona.* Tucson: University of Arizona Press.
Hogstad, O. 1975. Quantitative relations between hole-nesting and open-nesting species within a passerine breeding community. *Norwegian Journal of Zoology* 23:261–267.
Holechek, J. L., A. Tembo, A. Daniel, M. J. Fusco, and M. Cardenas. 1994. Long-term grazing influences on Chihuahuan Desert rangeland. *Southwestern Naturalist* 39:342–349.
Holmes, R. T., J. C. Schultz, and P. Nothnagle. 1979. Bird predation on forest insects: an exclosure experiment. *Science* 206:462–463.
Hooper, D. U., and P. M. Vitousek. 1998. Effects of plant composition and diversity on nutrient cycling. *Ecological Monographs* 68:121–149.
Hubbard, J. A., and G. R. McPherson. 1997. Acorn selection by Mexican jays: a test of a tri-trophic symbiotic relationship hypothesis. *Oecologia* 110:143–146.
Hubbard, J. P. 1977. The status of Cassin's sparrow in New Mexico and adjacent states. *American Birds* 31:933–941.
Humphrey, R. R. 1958. The desert grassland. *Botanical Review* 24:193–253.
———. 1970. *Arizona Range Grasses.* Tucson: University of Arizona Press.
———. 1987. *Ninety Years and 535 Miles: Vegetation Changes along the Mexican Border.* Albuquerque: University of New Mexico Press.

Huston, M. A. 1994. *Biological Diversity: The Coexistence of Species on Changing Landscapes.* Cambridge, Eng.: Cambridge University Press.

Jackson, R. B., O. E. Sala, J. M. Paruelo, and H. A. Mooney. 1998. Ecosystem water fluxes for two grasslands in elevated CO_2. *Oecologia* 113:537–546.

Jacobs, L. B. 1992. *Waste of the West: Public Lands Ranching.* Tucson: L. Jacobs.

Jepson-Innes, K., and C. E. Bock. 1989. Response of grasshoppers (Orthoptera: Acrididae) to livestock grazing in southeastern Arizona: differences between seasons and subfamilies. *Oecologia* 78:430–431.

Joern, A. 1992. Variable impact of avian predation on grasshopper assemblies in sandhills grassland. *Oikos* 64:458–463.

Joern, A., and K. H. Keeler, eds. 1995. *The Changing Prairie: North American Grasslands.* New York: Oxford University Press.

Johnson, R. R., and D. A. Jones, technical coordinators. 1977. *Importance, Preservation, and Management of Riparian Habitats: A Symposium.* USDA Forest Service, General Technical Report RM-43. Fort Collins, Colo.: Rocky Mountain Forest and Range Experiment Station.

Johnson, R. R., C. D. Ziebell, D. R. Patton, P. F. Ffolliott, and R. H. Hamre, technical coordinators. 1985. *Riparian Ecosystems and Their Management.* USDA Forest Service, General Technical Report RM-120. Fort Collins, Colo.: Rocky Mountain Forest and Range Experiment Station.

Johnson, W. C., C. S. Adkisson, T. R. Crow, and M. D. Dixon. 1997. Nut caching by blue jays (*Cyanocitta cristata* L.): implications for tree demography. *American Midland Naturalist* 138:357–370.

Jones, C., and R. W. Manning. 1996. The mammals. In *Rangeland Wildlife,* ed. P. R. Krausman, 29–38. Denver: Society for Range Management.

Jones, J. K., Jr., R. S. Hoffman, D. W. Rice, C. Jones, R. J. Baker, and M. D. Engstrom. 1992. *Revised Checklist of North American Mammals North of Mexico, 1991.* Occasional Paper No. 146. Lubbock: The Museum, Texas Tech University.

Jones, K. B. 1981. Effects of grazing on lizard abundance and diversity in western Arizona. *Southwestern Naturalist* 26:107–115.

———. 1988. Distribution and habitat associations of herpetofauna in Arizona: comparisons by habitat type. In *Management of Amphibians, Reptiles, and Small Mammals in North America,* R. C. Szaro, K. E. Severson, and D. R. Patton, technical coordinators, 109–128. USDA Forest Service, General Technical Report RM-166. Fort Collins, Colo.: Rocky Mountain Forest and Range Experiment Station.

Kareiva, P., and U. Wennergren. 1995. Connecting landscape patterns to ecosystem and population processes. *Nature* 273:299–302.

Kearney, T. H., and R. H. Peebles. 1960. *Arizona Flora.* Berkeley: University of California Press.

Keddy, P. 1989. *Competition.* London: Chapman & Hall.

Kellert, S. R., and E. O. Wilson, eds. 1993. *The Biophilia Hypothesis.* Washington, D.C.: Island Press.

Kenney, W. R., J. H. Bock, and C. E. Bock. 1986. Responses of the shrub, *Baccharis pteronioides,* to livestock exclosure in southeastern Arizona. *American Midland Naturalist* 116:429–431.

Keyser, A. J., G. E. Hill, and E. C. Soehren. 1998. Effects of forest fragment size, nest density, and proximity to edge on the risk of predation to ground-nesting passerine birds. *Conservation Biology* 12:986–994.

Kim, K. C., and R. D. Weaver, eds. 1994. *Biodiversity and Landscapes: A Paradox of Humanity.* Cambridge, Eng.: Cambridge University Press.

Knapp, A. K., J. M. Briggs, D. C. Hartnett, and S. C. Collins, eds. 1998. *Grassland Dynamics: Long-Term Ecological Research in Tallgrass Prairie.* Oxford: Oxford University Press.

Knight, R. L., and K. J. Gutzwiller, eds. 1995. *Wildlife and Recreationists: Coexistence through Management and Research.* Washington, D.C.: Island Press.

Knopf, F. L. 1994. Avian assemblages on altered grasslands. *Studies in Avian Biology,* no. 15: 247–257.

Knopf, F. L., R. R. Johnson, T. Rich, F. B. Samson, and R. C. Szaro. 1988. Conservation of riparian ecosystems in the United States. *Wilson Bulletin* 100:272–284.

Koenig, W. D. 1991. The effects of tannins and lipids on digestion of acorns by acorn woodpeckers. *Auk* 108:79–88.

Koenig, W. D., and S. H. Faeth. 1998. Effects of storage on tannin and protein content of cached acorns. *Southwestern Naturalist* 43:170–175.

Koenig, W. D., J. Haydock, and M. T. Stanback. 1998. Reproductive roles in the cooperatively breeding acorn woodpecker: incest avoidance versus reproductive competition. *American Naturalist* 151:243–255.

Koenig, W. D., and J. M. H. Knops. 1998. Scale of mast-seeding and tree-ring growth. *Nature* 396:225–226.

Koenig, W. D., and R. L. Mumme. 1987. *Population Ecology of the Cooperatively Breeding Acorn Woodpecker.* Princeton: Princeton University Press.

Koenig, W. D., R. L. Mumme, W. J. Carmen, and M. T. Stanback. 1994. Acorn production by oaks in central coastal California: variation within and among years. *Ecology* 75:99–109.

Korner, C., and F. A. Bazzaz, eds. 1996. *Carbon Dioxide, Populations, and Communities.* New York: Academic Press.

Kotler, B. P., J. S. Brown, and W. A. Mitchell. 1994. The role of predation in shaping the behaviour, morphology, and community organization of desert rodents. *Australian Journal of Zoology* 42:449–466.

Kozlowski, T. T., and C. E. Ahlgren, eds. 1974. *Fire and Ecosystems.* New York: Academic Press.

Krammes, J., ed. 1990. *Effects of Fire Management of Southwestern Natural Resources.* USDA Forest Service, General Technical Publication RM-91. Fort Collins, Colo.: Rocky Mountain Forest and Range Experiment Station.

Kramp, B. A., R. J. Ansley, and T. R. Tunnell. 1998. Survival of mesquite seedlings emerging from cattle and wildlife feces in a semi-arid rangeland. *Southwestern Naturalist* 43:300–312.

Krausman, P. R., ed. 1996. *Rangeland Wildlife.* Denver: Society for Range Management.

Krueper, D. J. 1994. Effects of land use practices on western riparian ecosystems. In *Status and Management of Neotropical Migratory Birds,* ed. D. M. Finch and P. W. Stengel, 321–328. USDA Forest Service, General Technical Report RM-229. Fort Collins, Colo.: Rocky Mountain Forest and Range Experiment Station.

Landres, P. B., J. Verner, and J. W. Thomas. 1988. Ecological uses of vertebrate indicator species: a critique. *Conservation Biology* 2:316–328.

Lauenroth, W. K., and W. A. Laycock, eds. 1989. *Secondary Succession and the Evaluation of Rangeland Condition.* Boulder, Colo.: Westview Press.

Leach, F. A. 1925. Communism in the California woodpecker. *Condor* 27:12–19.

Leopold, A. 1924. Grass, brush, and timber fire in southern Arizona. *Journal of Forestry* 22:1–10.

Leydet, F. 1988. *The Coyote: Defiant Songdog of the West.* Rev. ed. Norman: University of Oklahoma Press.

Licht, D. S. 1997. *Ecology and Economics of the Great Plains.* Lincoln: University of Nebraska Press.

Lima, S. L., and T. J. Valone. 1991. Predators and avian community organization: an experiment in a semi-desert grassland. *Oecologia* 86:105–112.

Limbaugh, R. H., III. 1992. *The Way Things Ought to Be.* New York: Pocket Books.

Limerick, P. N. 1987. *The Legacy of Conquest: The Unbroken Past of the American West.* New York: W. W. Norton.

Limerick, P. N. 1989. *Desert Passages: Encounters with the American Deserts.* Niwot: University Press of Colorado.

Lloyd, J., R. W. Mannan, S. Destefano, and C. Kirkpatrick. 1998. The effects of mesquite invasion on a southeastern Arizona grassland bird community. *Wilson Bulletin* 110:403–408.

Longland, W. S., and M. V. Price. 1991. Direct observations of owls and heteromyid rodents: can predation risk explain microhabitat use? *Ecology* 72:2261–2273.

Lonsdale, W. M. 1994. Inviting trouble: introduced pasture species in northern Australia. *Australian Journal of Ecology* 19:345–354.

Low, N., and B. Gleeson. 1998. *Justice, Society, and Nature: An Exploration of Political Ecology.* London: Routledge.

Luken, J. O., and J. W. Thieret, eds. 1997. *Assessment and Management of Plant Invasions.* New York: Springer-Verlag.

MacArthur, R. H. 1958. Population ecology of some warblers of northeastern coniferous forests. *Ecology* 39:599–619.

———. 1972. *Geographical Ecology.* New York: Harper & Row.

Mack, R. N., and J. N. Thompson. 1982. Evolution in steppe with few large hooved mammals. *American Naturalist* 119:757–773.

Maclean, N. 1976. *A River Runs through It, and Other Stories.* Chicago: University of Chicago Press.

Manning, R. 1995. *Grassland: The History, Biology, Politics, and Promise of the American Prairie.* New York: Viking Penguin.

Martin, S.C. 1975. *Ecology and Management of Southwestern Semidesert Grass-shrub Ranges: The Status of Our Knowledge.* Research Paper, USDA Forest Service, RM-156. Fort Collins, Colo.: Rocky Mountain Forest and Range Experiment Station.

Martin, T.E. 1995. Avian life history evolution in relation to nest sites, nest predation, and food. *Ecological Monographs* 65:101–127.

Mathies, T., and R.M. Andrews. 1995. Thermal and reproductive biology of high and low elevation populations of the lizard *Sceloporus scalaris:* implications for the evolution of viviparity. *Oecologia* 104:101–111.

Maurer, B.A. 1985. Avian community dynamics in desert grasslands: observational scale and hierarchical structure. *Ecological Monographs* 55:295–312.

McClaran, M.P. 1995. Desert grasslands and grasses. In *The Desert Grassland,* ed. M.P. McClaran and T.R. Van Devender, 1–30. Tucson: University of Arizona Press.

McClaran, M.P., and T.R. Van Devender, eds. 1995. *The Desert Grassland.* Tucson: University of Arizona Press.

McCullough, D.R., ed. 1996. *Metapopulations and Wildlife Conservation.* Washington, D.C.: Island Press.

McDonald, J.N. 1981. *North American Bison, Their Classification and Evolution.* Berkeley: University of California Press.

McKibben, B. 1998. *Maybe One.* New York: Simon & Schuster.

McLaughlin, S.P. 1992. Are floristic areas hierarchically arranged? *Journal of Biogeography* 19:21–32.

McPherson, G.R. 1995. The role of fire in desert grasslands. In *The Desert Grassland,* ed. M.P. McClaran and T.R. Van Devender, 130–151. Tucson: University of Arizona Press.

———. 1997. *Ecology and Management of North American Savannas.* Tucson: University of Arizona Press.

Meinzer, W. 1995. *Coyote.* Lubbock: Texas Tech University Press.

Mendelson, J.R., and W.B. Jennings. 1992. Shifts in the relative abundance of snakes in a desert grassland. *Journal of Herpetology* 26:38–45.

Merriam, C.H. 1898. *Life Zones and Crop Zones of the United States.* Division of Biological Survey, Bulletin No. 10. Washington, D.C.: U.S. Department of Agriculture.

Meserve, P.L., J.R. Gutierrez, J.A. Yunger, L.C. Contreras, and F.M. Jaksic. 1996. Role of biotic interactions in a small mammal assemblage in semiarid Chile. *Ecology* 77:133–148.

Milchunas, D.G., W.K. Lauenroth, and I.C. Burke. 1998. Livestock grazing: animal and plant biodiversity of shortgrass steppe and the relationship to ecosystem function. *Oikos* 83:65–74.

Milchunas, D.G., O.E. Sala, and W.K. Lauenroth. 1988. A generalized model of the effects of grazing by large herbivores on grassland community structure. *American Naturalist* 132:87–106.

Mitchell, J.C., and R.A. Beck. 1992. Free-ranging domestic cat predation on native vertebrates in rural and urban Virginia. *Virginia Journal of Science* 43:197–206.

Mönkkönen, M., P. Helle, and K. Soppela. 1990. Numerical and behavioural responses of migrant passerines to experimental manipulation of resident tits (*Parus* spp.): heterospecific attraction in northern breeding bird communities? *Oecologia* 85:218–225.
Mooney, H.A., and J.A. Drake, eds. 1986. *Ecology of Biological Invasions of North America and Hawaii.* New York: Springer-Verlag.
Murtaugh, P.A. 1996. The statistical evaluation of ecological indicators. *Ecological Applications* 6:132–139.
Naeser, R., and A. St. John. 1998. Water use and the future of the Sonoita Valley. In *The Future of Arid Grasslands: Identifying Issues, Seeking Solutions,* ed. B. Tellman, D.M. Finch, C. Edminster, and R. Hamre, 186–200. Proceedings RMRS-P-3. Fort Collins, Colo: USDA Forest Service, Rocky Mountain Research Station.
National Research Council. 1994. *Rangeland Health: New Methods to Classify, Inventory, and Monitor Rangelands.* Washington, D.C.: National Academy Press.
Neilson, R.P. 1986. High-resolution climatic analysis and Southwest biogeography. *Science* 232:27–34.
———. 1987. Biotic regionalization and climatic controls in western North America. *Vegetatio* 70:135–147.
Neilson, R.P., and D. Marks. 1995. A global perspective of regional vegetation and hydrologic sensitivities from climatic change. *Journal of Vegetation Science* 5:715–730.
Niemi, G.J., J.M. Hanowski, A.R. Lima, T. Nicholls, and N. Weiland. 1997. A critical analysis on the use of indicator species in management. *Journal of Wildlife Management* 61:1240–1252.
Noss, R.F. 1994. Cows and conservation biology. *Conservation Biology* 8:613–616.
Ohmart, R.D. 1994. The effects of human-induced changes on the avifauna of western riparian habitats. In *A Century of Avifaunal Change in Western North America,* ed. J.H. Jehl Jr. and N.K. Johnson, 273–285. Studies in Avian Biology, no. 15. Lawrence, Kans.: Cooper Ornithological Society.
Ortega, J.C. 1987a. Coyote food habits in southeastern Arizona. *Southwestern Naturalist* 32:152–155.
———. 1987b. Den site selection by the rock squirrel (*Spermophilus variegatus*) in southeastern Arizona. *Journal of Mammalogy* 68:792–798.
———. 1990a. Home-range size of adult rock squirrels (*Spermophilus variegatus*) in southeastern Arizona. *Journal of Mammalogy* 71:171–176.
———. 1990b. Reproductive biology of the rock squirrel (*Spermophilus variegatus*) in southeastern Arizona. *Journal of Mammalogy* 71:448–457.
———. 1991. The annual cycles of activity and weight of rock squirrels (*Spermophilus variegatus*) in southeastern Arizona. *American Midland Naturalist* 126:159–171.
Painter, E.L. 1995. Threats to the California flora: ungulate grazers and browsers. Madroño 42:180–188.
Painter, E.L., and A.J. Belsky. 1993. Application of herbivore optimization theory to rangelands of the western United States. *Ecological Applications* 3:2–9.

Pechmann, J. H., and D. B. Wake. 1997. Declines and disappearances of amphibian populations. In *Principles of Conservation Biology,* 2d ed., ed. G. K. Meffe and C. R. Carroll, 135–137. Sunderland, Mass.: Sinauer Associates.

Pfadt, R. E. 1994. *Field Guide to Common Western Grasshoppers.* Bulletin No. 912, Laramie: Wyoming Agricultural Experiment Station.

Phillips, A., J. Marshall, and G. Monson. 1964. *The Birds of Arizona.* Tucson: University of Arizona Press.

Polley, H. W. 1997. Implications of rising atmospheric carbon dioxide concentration for rangelands. *Journal of Range Management* 50:561–577.

Polley, H. W., H. S. Mayeux, H. B. Johnson, and C. R. Tischler. 1997. Viewpoint: atmospheric CO_2, soil water, and shrub/grass ratios in rangelands. *Journal of Range Management* 50:278–284.

Potter, C., ed. 1991. *Proceedings of a Workshop on the Impact of Cats on Native Wildlife.* Sydney: Endangered Species Unit, Australian National Parks and Wildlife Service.

Powell, A. N., and C. L. Collier. 1998. Reproductive success of Belding's Savannah sparrows in a highly fragmented landscape. *Auk* 115:508–513.

Price, J., S. Droege, and A. Price. 1995. *The Summer Atlas of North American Breeding Birds.* London: Academic Press.

Pulliam, H. R. 1975. Coexistence of sparrows: a test of community theory. *Science* 189:474–476.

———. 1983. Ecological community theory and the coexistence of sparrows. *Ecology* 64:45–52.

———. 1985. Foraging efficiency, resource partitioning, and the coexistence of sparrows. *Ecology* 66:1829–1836.

———. 1988. Sources, sinks, and population regulation. *American Naturalist* 132:652–661.

Pulliam, H. R., and M. R. Brand. 1975. The production and utilization of seeds in the plains grasslands of southeastern Arizona. *Ecology* 56:1158–1166.

Pulliam, H. R., and J. B. Dunning. 1987. The influence of food supply on local density and diversity of sparrows. *Ecology* 68:1009–1014.

Pulliam, H. R., and G. S. Mills. 1977. The use of space by wintering sparrows. *Ecology* 58:1393–1399.

Pyne, S., P. Andrews, and R. Laven. 1996. *Introduction to Wildland Fire.* 2d ed. New York: John Wiley.

Pysek, P., K. Prach, M. Rejmanek, and M. Wade, eds. 1995. *Plant Invasions: General Aspects and Special Problems.* Amsterdam: SPB Academic Publishers.

Reynolds, H. G. 1958. The ecology of the Merriam kangaroo rat (*Dipodomys merriami* Mearns) on the grazing lands of southern Arizona. *Ecological Monographs* 28:111–127.

Rice, J., B. W. Anderson, and R. D. Ohmart. 1984. Comparison of the importance of different habitat attributes to avian community organization. *Journal of Wildlife Management* 48:895–911.

Richards, R. T., J. C. Chambers, and C. Ross. 1998. Use of native plants on federal lands: policy and practice. *Journal of Range Management* 51:625–632.

Robbins, C. S., D. K. Dawson, and B. A. Dowell. 1989. Habitat area requirements of breeding birds of the Middle Atlantic States. *Wildlife Monographs* 103:1–34.
Robbins, C. S., J. R. Sauer, and B. G. Peterjohn. 1993. Population trends and management opportunities for neotropical migrants. In *Status and Management of Neotropical Migratory Birds,* ed. D. M. Finch and P. W. Stangel, 17–23. USDA Forest Service, General Technical Report RM-229. Fort Collins, Colo.: Rocky Mountain Forest and Range Experiment Station.
Robinett, D. 1992. Lehmann lovegrass and drought in southern Arizona. *Rangelands* 14:100–103.
Robinson, S. K. 1998. Another threat posed by forest fragmentation: reduced food supply. *Auk* 115:1–3.
Robinson, S. K., F. R. Thompson III, T. M. Donovan, D. R. Whitehead, and J. Faaborg. 1995. Regional forest fragmentation and the nesting success of migratory birds. *Science* 267:1987–1990.
Rolston, Holmes, III. 1994. *Conserving Natural Value.* New York: Columbia University Press.
Root, T. L. 1988. *Atlas of Wintering North American Birds.* Chicago: University of Chicago Press.
Rosenzweig, M. L. 1995. *Species Diversity in Space and Time.* Cambridge, Eng.: Cambridge University Press.
Rosenzweig, M. L., and J. Winakur. 1969. Population ecology of desert rodent communities: habitats and environmental complexity. *Ecology* 50:558–572.
Rotenberry, J. T., and J. A. Wiens. 1998. Foraging patch selection by shrubsteppe sparrows. *Ecology* 79:1160–1173.
Roundy, B. A., and S. H. Biedenbender. 1995. Revegetation in the desert grassland. In *The Desert Grassland,* ed. M. P. McClaran and T. R. Van Devender, 265–303. Tucson: University of Arizona Press.
Ruyle, G. B., B. A. Roundy, and J. R. Cox. 1988. Effects of burning on germinability of Lehmann lovegrass. *Journal of Range Management* 41:404–406.
Saab, V. A., C. E. Bock, T. D. Rich, and D. S. Dobkin. 1995. Livestock grazing effects in western North America. In *Ecology and Management of Neotropical Migratory Birds,* ed. T. E. Martin and D. M. Finch, 311–353. New York: Oxford University Press.
Sala, A., S. D. Smith, and D. A. Devitt. 1996. Water use by *Tamarix ramosissima* and associated phreatophytes in a Mojave Desert floodplain. *Ecological Applications* 6:888–898.
Samson, F. B., and F. L. Knopf, eds. 1996. *Prairie Conservation.* Washington, D.C.: Island Press.
Sanchini, P. J. 1981. Population structure and fecundity patterns in *Quercus emoryi* and *Q. arizonica* in southeastern Arizona. Ph.D. diss., University of Colorado, Boulder.
Savory, A., and J. Butterfield. 1999. *Holistic Management.* 2nd ed. Washington, D.C.: Island Press.
Schlesinger, W. H., J. F. Reynolds, G. L. Cunningham, L. F. Huenneke, W. M. Jarell, R. A. Virginia, and W. G. Whitford. 1990. Biological feedbacks in global desertification. *Science* 247:1043–1048.

Schluter, D., and R. R. Repasky. 1991. Worldwide limitation of finch densities by food and other factors. *Ecology* 72:1763–1774.

Schmitz, D. C., and D. Simberloff. 1997. Biological invasions: a growing threat. *Issues in Science and Technology* 13:33–40.

Schmutz, E. M., E. L. Smith, P. R. Ogden, M. L. Cox, J. O. Klemmedson, J. J. Norris, and L. C. Fiero. 1991. Desert grassland. In *Natural Grasslands: Introduction and Western Hemisphere,* ed. R. T. Coupland, 337–362. Ecosystems of the World 8A. Amsterdam: Elsevier.

Schnell, J. H. 1968. The limiting effects of natural predation on experimental cotton rat populations. *Journal of Wildlife Management* 32:698–711.

Schoener, T. W. 1982. The controversy over interspecific competition. *American Scientist* 70:586–595.

———. 1983. Field experiments on interspecific competition. *American Naturalist* 122:240–285.

Seastedt, T. R., and A. K. Knapp. 1993. Consequences of nonequilibrium resource availability across multiple time scales: the transient maxima hypothesis. *American Naturalist* 141:621–633.

Sih, A., P. Crowley, M. McPeek, J. Petranka, and K. Strohmeier. 1985. Predation, competition, and prey communities: a review of field experiments. *Annual Review of Ecology and Systematics* 16:269–311.

Skagen, S. K., C. P. Melcher, W. H. Howe, and F. L. Knopf. 1998. Comparative use of riparian corridors and oases by migrating birds in southeast Arizona. *Conservation Biology* 12:896–909.

Smith, H. M., D. Chiszar, A. Chiszar, D. L. Auth, J. Auth, C. Henke, C. E. Bock, J. H. Bock, J. A. Rybak, R. L. Holland, K. Bonine, and G. J. Watkins-Colwell. 1998. Slevin's bunch grass lizard (*Sceloporus slevini*) decimated on the Sonoita Plain, Arizona. *Herpetological Review* 29:225–226.

Smith, H. M., G. J. Watkins-Colwell, E. A. Liner, and D. Chiszar. 1996. *Sceloporus scalaris auctorum* a super species (Reptilia: Sauria). *Bulletin of the Maryland Herpetological Society* 32:70–74.

Snyder, G. 1995. *A Place in Space: Ethics, Aesthetics, and Watersheds.* Washington, D.C.: Counterpoint.

Soulé, M. E., ed. 1986. *Conservation Biology: The Science of Scarcity and Diversity.* Sunderland, Mass.: Sinauer Associates.

Sowls, L. K. 1984. *The Peccaries.* Tucson: University of Arizona Press.

Stacey, P. B., and C. E. Bock. 1978. Social plasticity in the acorn woodpecker. *Science* 202:1298–1300.

Stacey, P. B., and W. D. Koenig. 1984. Cooperative breeding in the acorn woodpecker. *Scientific American* 251:114–121.

———, eds. 1990. *Cooperative Breeding in Birds.* Cambridge, Eng.: Cambridge University Press.

Stamp, N. E. 1978. Breeding birds of riparian woodlands in south-central Arizona. *Condor* 80:64–71.

Stapp, P. 1997. Community structure of shortgrass-prairie rodents: competition or risk of intraguild predation? *Ecology* 78:1519–1530.

Stebbins, R. C. 1985. *A Field Guide to Western Reptiles and Amphibians.* 2d ed. Boston: Houghton Mifflin.

Steele, M. A., T. Knowles, K. Bridle, and E. L. Simms. 1993. Tannins and partial consumption of acorns: implications for dispersal of oaks by seed predators. *American Midland Naturalist* 130:229–238.

Stromberg, J. C., R. Tiller, and B. Richter. 1996. Effects of groundwater decline on riparian vegetation of semiarid regions: the San Pedro, Arizona. *Ecological Applications* 6:113–131.

Stromberg, M. R. 1990. Habitat, movements, and roost characteristics of Montezuma quail in southeastern Arizona. *Condor* 92:229–236.

Strong, D. R., Jr. 1983. Natural variability and the manifold mechanisms of ecological communities. *American Naturalist* 122:636–660.

Strong, T. R., and C. E. Bock. 1990. Bird species distribution patterns in riparian habitats in southeastern Arizona. *Condor* 92:866–885.

Szaro, R. C., and M. D. Jakle. 1985. Avian use of a desert riparian island and its adjacent scrub habitat. *Condor* 87:511–519.

Taylor, C. A., Jr., T. D. Brooks, and N. E. Garza. 1993. Effects of short duration and high-intensity, low-frequency grazing systems on forage production and composition. *Journal of Range Management* 46:118–121.

Tellman, B. 1998. Stowaways and invited guests: how some exotic plant species reached the American Southwest. In *The Future of Arid Grasslands: Identifying Issues, Seeking Solutions,* ed. B. Tellman, D. M. Finch, C. Edminster, and R. Hamre, 144–149. Proceedings RMRS-P-3. Fort Collins, Colo.: USDA Forest Service, Rocky Mountain Research Station.

Terbourgh, J. 1989. *Where Have All the Birds Gone?* Princeton: Princeton University Press.

Thomas, P. A., and P. Goodson. 1992. Conservation of succulents in desert grasslands managed by fire. *Biological Conservation* 60:91–100.

Tiedemann, A. R., and J. O. Klemmedson. 1973. Nutrient availability in desert grassland soils under mesquite (*Prosopis juliflora*) trees and adjacent open areas. *Soil Society of America Proceedings* 37:107–111.

Tracy, K. N., D. M. Golden, and T. O. Crist. 1998. The spatial distribution of termite activity in grazed and ungrazed Chihuahuan Desert grassland. *Journal of Arid Environments* 40:77–89.

Truett, J. 1996. Bison and elk in the American Southwest: in search of the pristine. *Environmental Management* 20:195–206.

Turner, R. M., L. H. Applegate, R. M. Bergthold, S. Gallizioli, and S. C. Martin. 1980. *Arizona Range Reference Areas.* USDA Forest Service, General Technical Report RM-79. Fort Collins, Colo.: Rocky Mountain Research Station.

Turner, R. M., J. E. Bowers, and T. L. Burgess. 1995. *Sonoran Desert Plants, an Ecological Analysis.* Tucson: University of Arizona Press.

Valone, T. J., and J. H. Brown. 1995. Effects of competition, colonization, and extinction on rodent species diversity. *Science* 267:880–883.
Van Devender, T. R. 1995. Desert grassland history: changing climates, evolution, biogeography, and community dynamics. In *The Desert Grassland,* ed. M. P. McClaran and T. R. Van Devender, 68–99. Tucson: University of Arizona Press.
Vandermeer, J. 1996. *Reconstructing Biology: Genetics and Ecology in the New World Order.* New York: John Wiley.
Vavra, M., W. A. Laycock, and R. D. Pieper, eds. 1994. *Ecological Implications of Livestock Herbivory in the West.* Denver: Society for Range Management.
Vickery, P. D. 1996. Grasshopper sparrow. In *The Birds of North America,* No. 239, ed. A. Poole and F. Gill. Philadelphia: Academy of Natural Sciences; Washington, D.C.: American Ornithologists' Union.
Vickery, P. D., M. L. Hunter, Jr., and S. M. Melvin. 1994. Effects of habitat area on the distribution of grassland birds in Maine. *Conservation Biology* 8:1087–1097.
Vinton, M. A., D. C. Hartnett, E. J. Finck, and J. M. Briggs. 1993. The interactive effects of fire, bison (*Bison bison*) grazing and plant community composition in tallgrass prairie. *American Midland Naturalist* 129:10–18.
Wake, D. B. 1998. Action on amphibians. *Trends in Ecology and Evolution* 13:379–380.
Walker, B. H., D. Ludwig, C. S. Holling, and R. M. Peterman. 1981. Stability of semiarid savanna grazing systems. *Journal of Ecology* 69:473–498.
Waters, J. R., B. R. Noon, and J. Verner. 1990. Lack of nest site limitation in a cavity-nesting bird community. *Journal of Wildlife Management* 54:239–245.
Webb, E. A., and C. E. Bock. 1990. Relationship of the Botteri's sparrow to sacaton grassland in southeastern Arizona. In *Managing Wildlife in the Southwest,* ed. P. R. Krausman and N. S. Smith, 199–209. Tucson: Arizona Chapter of the Wildlife Society.
———. 1996. Botteri's sparrow (*Aimophila botterii*). In *The Birds of North America,* No. 216, ed. A. Poole and F. Gill. Philadelphia: Academy of Natural Sciences; Washington, D.C.: American Ornithologists' Union.
Weltz, M., and M. K. Wood. 1986. Short-duration grazing in central New Mexico: effects on infiltration rates. *Journal of Range Management* 39:365–368.
Weltzin, J. F., S. Archer, and R. K. Heitschmidt. 1997. Small-mammal regulation of vegetation structure in a temperate savanna. *Ecology* 78:751–763.
Wenz, P. 1998. *Environmental Justice.* Albany: State University of New York Press.
White, M. R., R. D. Pieper, G. B. Donart, and L. W. Trifaro. 1991. Vegetation response to short-duration and continuous grazing in south-central New Mexico. *Journal of Range Management* 44:399–403.
White, S. 1974. Seed and vegetation patterns with respect to grazing on a southern Arizona rangeland. M.S. thesis, Arizona State University, Tempe.
Whitford, W. G. 1976. Temporal fluctuations in density and diversity of desert rodent populations. *Journal of Mammalogy* 57:351–369.
———. 1997. Desertification and animal biodiversity in the desert grasslands of North America. *Journal of Arid Environments* 37:709–720.

Whittaker, R. H. 1975. *Communities and Ecosystems.* 2d ed. New York: Macmillan.

Wiegert, R. G. 1972. Avian versus mammalian predation on a population of cotton rats. *Journal of Wildlife Management* 36:1322–1327.

Wiens, J. A. 1983. Avian community ecology: an iconoclastic view. In *Perspectives in Ornithology,* ed. A. H. Brush and G. A. Clark Jr., 355–403. Cambridge, Eng.: Cambridge University Press.

———. 1989. *The Ecology of Bird Communities.* 2 vols. Cambridge, Eng.: Cambridge University Press.

Wiens, J. A., R. G. Cates, J. T. Rotenberry, N. Cobb, B. Van Horne, and R. A. Redak. 1991. Arthropod dynamics on sagebrush (*Artemisia tridentata*): effects of plant chemistry and avian predation. *Ecological Monographs* 61:299–321.

Wiens, J. A., and J. T. Rotenberry. 1981a. Habitat associations and community structure of birds in shrubsteppe environments. *Ecological Monographs* 51:21–41.

———. 1981b. Morphological size ratios and competition in ecological communities. *American Naturalist* 117:592–599.

Wilcove, D. S. 1985. Nest predation in forest tracts and the decline of migratory songbirds. *Ecology* 66:1211–1214.

Wilson, E. O. 1992. *The Diversity of Life.* New York: W. W. Norton.

Wilson, J. P. 1995. *Islands in the Desert: A History of the Uplands of Southeastern Arizona.* Albuquerque: University of New Mexico Press.

Windberg, L. A., S. M. Ebbert, and B. T. Kelly. 1997. Population characteristics of coyotes (*Canis latrans*) in the northern Chihuahuan Desert of New Mexico. *American Midland Naturalist* 138:197–207.

Wright, H. A., and A. W. Bailey. 1982. *Fire Ecology, United States and Southern Canada.* New York: John Wiley.

Wuerthner, G. 1994. Subdivisions vs. agriculture. *Conservation Biology* 8:905–908.

INDEX

Text: Adobe Garamond
Display: Perpetua and Adobe Garamond
Composition: Impressions Book and Journal Services, Inc.
Printing and binding: Edwards Brothers, Inc.
Maps: Bill Nelson